# NANOTECHNOLOGY

*Legal Aspects*

# PERSPECTIVES IN NANOTECHNOLOGY

## Series Editor
### Gabor L. Hornyak

# NANOTECHNOLOGY

## *Legal Aspects*

Patrick M. Boucher

CRC Press
Taylor & Francis Group
Boca Raton   London   New York

CRC Press is an imprint of the
Taylor & Francis Group, an **Informa** business

CRC Press
Taylor & Francis Group
6000 Broken Sound Parkway NW, Suite 300
Boca Raton, FL 33487-2742

© 2008 by Taylor & Francis Group, LLC
CRC Press is an imprint of Taylor & Francis Group, an Informa business

International Standard Book Number-13: 978-1-4200-5347-0 (Softcover)

---

**Library of Congress Cataloging-in-Publication Data**

---

Boucher, Patrick M.
    Nanotechnology : legal aspects / Patrick M. Boucher.
      p. cm.
    Includes bibliographical references and index.
    ISBN 978-1-4200-5347-0 (alk. paper)
    1. Nanostructured materials industry--Law and legislation. 2. Nanostructured materials industry--Law and legislation--United States. I. Title.

K3924.H54B66 2008
343'.0786205--dc22                                2007037938

---

**Visit the Taylor & Francis Web site at**
**http://www.taylorandfrancis.com**

**and the CRC Press Web site at**
**http://www.crcpress.com**

*For Dad*

# Contents

# Series Foreword

Welcome to the Perspectives in Nanotechnology Series—a group of short, readable paperback books dedicated to expanding your knowledge about a new and exciting technology. The book you are about to read involves subject matter that goes beyond the laboratory and the production line. It is not about technical details—the book you have taken on board your connecting flight, commuter train or bus or to your hotel room involves a specific aspect of nanotechnology that will have some impact on your life, the welfare of your family and the wealth and security of this nation. The degree of this impact may be unnoticeable, slight, overwhelming or any place in between those extremes depending on the specific application, its magnitude and the scope of its distribution. Those of us who are able to recognize trends, conduct efficient research, plan ahead and adapt will succeed in a new world enhanced by nanotechnology. This book in the Perspectives in Nanotechnology Series hopefully will act as the catalyst for your *fantastic journey*.

Each book in the series focuses on a selected aspect of nanotechnology. No technology exists in a vacuum. All technology is framed within the contexts of societal interactions, laws and practices. Once a technology is introduced to a society, the society must deal with it. The impact of a technology on culture, politics, education and economics depends on many complex factors—just reflect for a moment on the consequences (good and bad) of the computer, the automobile or the atomic bomb. Nanotechnology is designated to be the "next industrial revolution." Although there is much hype associated with nanotechnology, the ability to manipulate atoms and molecules in order to fabricate new materials and devices that possess remarkable properties and functions alone should be enough of a hook to draw you in.

The impact of new technology is more relevant than ever. Consider that our world is highly integrated, communication occurs instantaneously and that powerful geopolitical and economic pressures are in the process of continually changing the global landscape. We repeat—the degree of the impact of nanotechnology may be unnoticeable, slight, overwhelming or anyplace in between. Those of us who are able to recognize trends, conduct efficient research, plan ahead and adapt will succeed. It is all about survival. It always has been. Darlene Geis in her book, *Dinosaurs and Other Prehistoric Animals*, states:

> ...and finally even the mighty T-Rex died out, too. His size and strength and remarkable jaws were of no use to him in a world that was changing and where his food supply was slowly disappearing. In the end, the king was no greater than his subjects in a world whose rule has always been Change with Me—or Perish![1]

Although stated with a bit of drama, the quotation does bring the point across quite effectively. Your future is in your hands—perhaps holding this very book.

---

## Societal Implications

Societal aspects (implications) consist of a broad family of highly integrated components and forces that merge with technology to form our civilization. Government, business, academia and other social institutions have evolved over millennia and are in a constant state of dynamic flux. Civilizations change for many reasons. Technology always has been one of the primary drivers of this change. The change may be beneficial, detrimental or anywhere in between. From the first stone implement, the iron of the Hittites to the microchip, technology has always played a major role in the shaping of society. Societal implications of nanotechnology are rooted in the technology. Societal implications in turn have the capacity to alter any technology. How many times have social forces inspired a new technology? The technology developed in the space program is one example of such a relationship—the development of penicillin another.

What exactly are "societal implications"? How do they relate to nanotechnology? In this series, we intend to cover a wide variety of topics. The societal implications of nanotechnology are both numerous and diverse and encompass the legal, ethical, cultural, medical, and environmental disciplines. National security, education, workforce development, economic policy, public policy, public perception, and regulation are but a few of the areas we plan to address in the near future.[2] All aspects of government, business and academia are subject to the influence of nanotechnology. All vertical industrial sectors will be impacted by nanotechnology—aerospace, health care, transportation, electronics and computing, telecommunications, biotechnology, agriculture, construction and energy. For example, all Fortune 500 companies already have staked a claim in nanotechnology-based products. Service industries that focus on intellectual property and technology transfer, health and safety, environmental management and consulting, workforce sourcing and job placement, education development and curriculum, and investment and trading already engage the challenges brought about by nanotechnology. There is no lack of subject matter. We plan to cover the most urgent, the most relevant, and the most interesting topics.

Ethical implications are associated with every form of technology. Artificial intelligence, weapon systems, life-extending drugs, surveillance, altered organisms and social justice all have built-in moral implications—ready for us to discuss. Nanotechnology is creating new ethical dilemmas while simultaneously exacerbating (or alleviating) older ones. Nanotechnology is already changing our legal system. How does one go about obtaining a patent of a process or material that is the result of an interdisciplinary collaboration, e.g., the convergence of engineering, chemistry, physics and biology? Even more so, the environmental footprint of nanotechnology is expected to be three orders of magnitude less than that of any current technology. The health (and environmental) consequences of nanomaterials are mostly unknown. And what of public perception? How many of you want a nanotech research center in your back yard (are you a NIMBY)? How should we update our educational system to accommodate nanotechnological topics? What should we do to make sure our workforce is current and prepared? How will your job or career be influenced by nanotechnology?

There are other relevant questions. How does one go about building a nanobusiness? What new kinds of partnerships are required to start a business and what exactly is the *barrier of entry* for such an undertaking? What are *nanoeconomic clusters*? What Fortune 500 companies and what business sectors require a book in this Series to describe its NT profile? And what of investing and funding? What is the status of nanotechnology programs on the international stage? What about nanotechnology and religion? What about the *future of nanotechnology*? The list goes on.

## The Books

Web resources that address societal implications of nanotechnology are plentiful but usually offer encapsulated or cursory information. On the other hand, comprehensive (but tedious) summary reports produced by research and marketing firms are suitable for the serious investor but require a major financial commitment to procure and therefore, are generally not available to the public at large. In addition, government entities, e.g., the National Nanotechnology Initiative (http://www.nano.gov), have generated comprehensive reports on the societal impact of NT.[1] Such documents, although excellent, are generally not well known to the public sector. A reader-friendly, affordable book with commercial appeal that targets the nano-aware (as well as the unaware) layperson or expert in the field offers a convenient alternative to the options listed above.

The intent of each book is to be informative, compelling and relevant. The books, in general, adhere to the criteria listed below.

- **Readability**. Each book is 200 to 300 pages long, with easy-to-read font and is abundant with non-technical but certainly non-ponderous language.
- **References**. Each book is well researched and provides links to more detailed sources when required.

- **Economical Pricing**. Each book is priced within easy reach and designed for accelerated distribution at conferences and other venues.
- **Subject Matter**. The subject of each book is relevant to nanotechnology and represents the cutting-edge in the state-of-the-art.
- **Relevance**. The books are dynamic. We must stay current if we are to abide by T-Rex's rule! Specifically, the content will stay relevant in the form of future editions as the climate of nanotechnology is expected to change dynamically over the years to come. A strong temporal component is inherent in the Perspectives in Nanotechnology Series.

It is our hope that readers delve into a book about their special interest but also to transform themselves into a state of *nano-readiness*. Are you nano-ready? Do you want to be able to recognize the drivers that surround nanotechnology and its potential promise? Do you want to be able to learn about the science, technology and potential implications? Are ready at this time to plan and adapt to changes? Do you want to become an agent of change? Do you want success in that future? If your answers are, in order—NO, YES, YES, NO, YES and YES—you are ready to begin reading this book.

**Gabor L. Hornyak**
*Series Editor*

## References

1. D. Geis, *Dinosaurs and Other Prehistoric Animals*, Grosset & Dunlap, New York (1959).
2. M.C. Roco and W.S. Bainbridge, Eds., *Societal Implications of Nanoscience and Nanotechnology*, National Science Foundation, Arlington, Virginia (2001).
3. M.C. Roco and W.S. Bainbridge, Eds., *Nanotechnology: Societal Implications—Maximizing Benefits for Humanity*, Report of the National Nanotechnology Initiative Workshop, December 2–3 (2003).

# *Author*

Patrick M. Boucher holds a Ph.D. in physics (Queen's University, Canada) and a J.D. (Touro College, United States). His technical publications have been in the areas of condensed matter, nuclear, and astrophysics. For several years, he was associated with *Physical Review B*, which publishes much of the world's technical nanotechnology research and where he acted as associate editor and managed the journal's scientific editorial staff. He is currently a patent attorney practicing in Denver, Colorado, as a partner of Townsend and Townsend and Crew LLP.

# Introduction

In April 2006, Magic Nano, a commercial product designed to render glass and ceramic surfaces repellant to dirt and water, was recalled by the German government because of reports of respiratory problems experienced by some seventy people during a one-week period in March 2006. In July 2006, Elan Pharmaceuticals filed a lawsuit against Abraxis Bioscience, Inc., alleging that the drug Abraxane, a nanoparticle formulation of paclitaxel marketed by Abraxis for the treatment of metastatic breast cancer, infringed patents owned by Elan. Just before Christmas 2006, demonstration of surface conduction electron emitter displays—a new alternative to LCD and plasma displays based on nanotechnology—was prevented by a lawsuit filed by Nano-Proprietary against Canon.

These are just some of the legal issues surrounding nanotechnology that have begun to surface in the news. In many ways, nanotechnology has so far been surprisingly immune to entanglement in public legal issues, although there is no question that prudent nanotechnology companies have had legal advice to position themselves for addressing a variety of issues. But for the most part, researchers have so far been able to conduct their investigations with relatively little concern about various legal ramifications.

Such a circumstance is unlikely to continue. As nanotechnology continues to be developed and as financial interests become more important, legal doctrines are increasingly likely to play an important role. Particularly as nanotechnology continues to be used in commercial products, private parties will use the mechanisms that the law provides to try to develop and exploit financial advantages. Governments will also increasingly become involved with integrating nanotechnology into a broader legal framework that allows for a variety of different concerns to be addressed. It is already the case that existing laws have a generality that permits them to be applied to nanotechnology and the illustrations above represent just a small example of ways in which those existing laws are actively being used. It is also certain that as nanotechnology becomes more pervasively included in products, governments will feel compelled to generate legislation targeted to issues specific to nanotechnology.

This book provides an analysis of some of the legal issues that surround nanotechnology. At its most fundamental level, the recognized purpose of laws in society is to control people's behavior. The manner in which laws affect such control varies depending on circumstances and may sometimes be manifested as incentives designed to encourage people to engage in activity that society views as beneficial. The converse of such an approach is realized in laws whose purpose is instead to punish those who engage in activity that society deems unacceptable or otherwise undesirable.

Nanotechnology provides a cogent example of a discipline that is subject to a wide range of laws including those that act as a carrot to incentivize desirable behavior and those that act as a stick to discourage unacceptable activities. The spectrum of laws classified in this way approximates a lifecycle that might be assigned to a nanotechnology product. At one end of the spectrum are laws that may be used to encourage researchers to apply their intellectual abilities to developing the science and engineering used in producing nanotechnology products. These laws include mechanisms for recognizing that what these researchers develop is their "intellectual property" and that this has value. The law provides a significant number of rights depending on the character of what is developed.

At the other end of the spectrum are laws that are intended to hold individuals responsible for their actions, either by punishing them for engaging in behavior viewed as contrary to the objectives of society or by requiring them to compensate those whom their behavior harms. Such laws may come into play when a product that incorporates nanotechnology is used in a manner that causes harm to individuals. A number of issues are relevant to assigning liability for the harm in such circumstances, with criminal laws focusing on punishment by the state and civil laws mostly focusing on providing private compensation to injured parties.

Intermediate in this spectrum are a variety of regulatory laws. The regulation of activities reflects an attempt by society to act preemptively to prevent certain results from occurring. In this way, such laws differ from those that are specifically designed to encourage behavior or are specifically designed to punish behavior. This difference in approach is reflected in the way in which most regulatory restrictions are implemented. Rather than enact specific laws to govern activities in this preemptive way, lawmakers create a regulatory body that has expertise in the relevant area and delegate a portion of their power to that body. The regulatory body then generally has a more direct interaction with the community that is most affected by limitations on its activities, providing mechanisms to solicit input from that community before promulgating restrictive rules.

This book is organized in three parts that address different portions of this legal spectrum. The first part, "Protection," describes mechanisms by which creative and inventive parties may obtain control over their developments, particularly in a manner that allows those developments to be exploited for financial gain. This first part discusses how developers of nanotechnology products may obtain and enforce patents, copyrights, and other forms of intellectual property. The second part, "Regulation," provides an overview of some of the existing regulatory bodies that will impact the practical use of nanotechnology. By no means are all bodies that could potentially regulate nanotechnology discussed in this section; rather the section uses illustrations of a number of different regulations to demonstrate different kinds of approaches that might be taken in regulating nanotechnology. The final part of the book, "Liability," focuses on a variety of different legal doctrines that may be used to hold parties responsible for their actions. While the greatest bulk of that part addresses negligence and product-liability doctrines that have their basis in civil law, some discussion is also provided of the potential impact of criminal-law principles to nanotechnology.

The discussion of these principles is intended to be relatively general throughout the book, but it is necessary in many instances to provide at least some specific details as to how the law applies. When this is done, I have used U.S. law, which is the body of law with which I am most familiar. This should not detract significantly from the main thrust of the book, which is to illustrate how nanotechnology is affected by different legal principles. In particular, this book is not intended to be a legal treatise. In many instances, I have simplified some of the subtleties of certain legal doctrines or ignored some exceptions that could apply in order to bring out the concepts I am hoping to illustrate more plainly. The book should therefore not be relied upon to provide the answer to any specific legal question. A competent attorney who is aware of case-law developments that have occurred since this book was written and who is familiar with the precise details and exceptions of the relevant legal doctrines is essential.

Even though U.S. law has been used for illustrative purposes, it is worth remarking that even such a concept is not necessarily well defined. The United States is structured as a federal system in which a large number of states are considered fundamentally to be independently sovereign. They have yielded some of this sovereignty to the national (or federal) government, which is able to legislate in certain areas, but have retained significant power for themselves. In this kind of context, discussions of legal principles are made in a generalized fashion, with the recognition that each jurisdiction may include its own unique variants to the generally prevailing views. In many respects, the same kinds of differences exist when considering the laws of other nations. While there are certainly variations among different countries—and while the variants in an international context may well be larger than the variants within a single country like the United States—most of the generalized principles still apply. The simple fact is that the same kinds of legal issues need to be addressed in all countries and, for the most part, similar principles are developed.

Notably, this book is one of a series of texts that discuss the societal implications of nanotechnology. As Louis Hornyak notes in his foreword, the series of books attempts to provide some perspective on these societal impacts, with a broader outlook being available by considering the ways in which different disciplines intersect and provide different viewpoints of similar issues. In some respects, it is not a trivial task to demarcate the separate subjects that are addressed in the series, and there is inevitably some overlap between various of the books.

The reader is encouraged to consult the other volumes in this series when his or her own research encounters these interfaces. For example, the protection of intellectual property that results from the scientific and engineering development of nanotechnology is an important consideration in the structuring of nanotechnology businesses. In many instances, such intellectual property is the only really valuable asset that an early-stage company has. Other considerations important to the structuring of such businesses are discussed in detail in the companion book on business aspects of nano-technology by Michael Burke. Among other things, the presentation in that volume places the intellectual property considerations within the broader context of issues that must be faced in successfully developing a nanotech-nology business. The volume on education and workforce development by Louis Hornyak also integrates well with these issues by explaining concerns related to ensuring that personnel will be adequately trained to develop such intellectual property and to make real contributions to nanotechnology businesses.

And while the section on regulation in this book includes a discussion of various mechanisms to address environmental concerns, there is relatively little specific discussion of what those concerns are or how to assess their risk realistically. Those topics are instead covered in the companion book by Jo Anne Shatkin on nanotechnology and the environment. A much more realistic understanding of the societal context of the interplay between envi-ronmental and nanotechnology issues results from the combination of the books, with the companion book providing information useful in discerning *what* should be done and this book on legal aspects describing mechanisms for *how* it should be accomplished. The same is true with ethical issues, which are discussed in considerable detail in the book in this series by Deb Bennett-Woods. That book provides a framework by which ethical issues raised by nanotechnology can be understood and evaluated. This book on legal issues describes a number of mechanisms that may give rise to solutions to those ethical problems.

It is also worth noting that the discussion provided in this book is in a cer-tain sense static. While legal principles have developed a certain flexibility to account for the persistence of change, they seem at any given moment to be relatively fixed. But nanotechnology remains in its infancy and is certain to mature in ways that will challenge the legal framework as it exists today. Aspects of what principles will govern how nanotechnology will evolve over time are developed in the book on the future of nanotechnology by Thomas J. Frey. In many ways, an accurate understanding of these principles may aid in the refinement of laws as they exist today.

One of the goals of representative systems of government, regardless of the details of their specific structure, is to provide a mechanism that allows for coherent decisions to be made in a collective fashion. In such systems, the views of one individual or group are of almost no consequence unless they have the power to persuade others. And the process of attempting to per-suade others of the legitimacy of a certain point of view naturally accounts for the concerns of contrary views. In this way, societies aspire to act in a way that is more responsible, more informed, and more ethical than is the sum of its individual parts. In a similar manner—albeit on a much smaller scale—it is hoped that this book will contribute to the Perspectives in Nanotechnology series in a way that permits greater insight into how legal issues bear on nanotechnology than would be possible with a volume divorced from that wider context.

---

## Note

1. *Washington Post*, "Nanotech Product Recalled in Germany," April 6, 2006, p. A02. Later investigations confirmed that despite the name of the product, Magic Nano did not actually contain nanoparticles.

# I

## Protection

### A. Patents

#### 1. Introduction

Filippo Brunelleschi is best known for his astonishing design of the dome of Santa Maria del Fiore Cathedral in the heart of Florence (Figure I.1). Built primarily between the twelfth and fourteenth centuries, the dome held the record as the largest dome in the world until 1928 when construction of the Leipzig Market dome was completed; the record was thus only passed with the development of a completely new technology, in this case the use of reinforced concrete instead of traditional masonry techniques for building domes. Even now the dome of Santa Maria del Fiore remains larger than the dome of the Capitol Building in Washington DC, larger than the dome at St Paul's Cathedral in London, larger than the dome in the Pantheon in Rome, and even larger than the dome in St. Peter's Basilica in the Vatican. The story of how the dome was built, especially within the context of Brunelleschi's nearly lifelong competition with Lorenzo Ghiberti, is a remarkable story that highlights the achievements possible by those possessed with unusual ingenuity.[1]

What is somewhat less well known about Brunelleschi is that he was also the first person in history to be awarded a patent for a technical invention. Brunelleschi had developed a technique for transporting marble upstream the Arno River to Florence from the quarries at Carrara. During the height of the Italian Renaissance period, the transportation of marble for use in creating sculptural works of art was both difficult and costly. Brunelleschi refused to disclose his technique, fearing that others would make use of his insight without him receiving any direct benefit. The Republic of Florence accordingly granted a patent to Brunelleschi in 1421, giving him a monopoly on the manufacture of his invention, a barge that included hoisting gear to facilitate transportation of the marble. The monopoly lasted three years and permitted Brunelleschi to benefit from inventing what his contemporaries described as *"Il Badalone"* ("The Monster").

In its grant of the patent to Brunelleschi, the Florentine Republic included a preamble that expressed the basic pact that governments continue to make with inventors in order to have them disclose their inventions:

**FIGURE I.1**
A view of Santa Maria del Fiore Cathedral showing the design of Brunelleschi's dome.

> The admirable Filippo Brunelleschi, a man of the most perspicacious
> intellect, industry and invention, citizen of Florence, has invented some
> machine or kind of ship, by means of which he thinks he can easily, at
> any time, bring in any merchandise and load on the river Arno and on
> any other river or water, for less money than usual, and with several
> other benefits to merchants.... [H]e refuses to make such machine avail-
> able to the public, in order that the fruit of his genius and skill may not be
> reaped by another without his will and consent.... [I]f he enjoyed some
> prerogative concerning this, he would open up what he is hiding and
> would disclose it to all.

Since the time of Brunelleschi, virtually every government has adopted
some form of patent system, in which the government grants an inventor
certain benefits in exchange for the public disclosure of his or her invention.
Almost always, these benefits take the form of rights backed by the legal
authority of the state that permit the inventor to control use of the invention.

### i. Monopoly Powers

This basic agreement between inventors and governments remains at the heart of all patent systems to this day. By inducing inventors to disclose their inventions, governments increase the baseline of public knowledge, enabling other inventors to improve on those inventions in ways that were not conceived of by the original inventor. In this way, the system continually accelerates the pace of innovation by progressively making innovations accessible to as many people as possible.

Ever since the creation of patent systems, the monopoly powers that have been granted to inventors have, at the same time, caused consternation among those who wish to use the inventions. This is particularly true when the monopoly is being exercised at a time further removed from the original invention. When a useful invention is first disclosed, people are generally willing to acknowledge the creativity it represents and to concede that the inventor is entitled to some reward for having conceived of such a useful idea and publicly shared it. But over time, the invention becomes well known and common, and this initial perspective becomes eroded. People begin to bemoan the fact that, rather than spur their own innovation, the monopoly power is acting to prevent them from building on the original invention because they must secure the inventor's permission to be able to use it.

This pattern has always existed and is very much present today as these same kinds of criticisms continue to be made. In some cases, the criticisms (of the past or present) may be valid, with the monopoly power granted to the inventor being too strong or persisting for too great a time. And what was an effective compromise to achieve the goals of spurring innovation in the fifteenth century may prove not to be the best compromise in later centuries—the technologies themselves are different, as reflected in the way innovations are made; the context of technology in society as a whole is different; and certain technologies may be viewed as so fundamental to basic human existence that monopolies on those technologies are viewed with public suspicion or distaste.

For this reason, governments continually reconsider the specifics of the patent system. They modify the definition of what types of technology may be patentable, and different governments frequently develop different answers. This is perhaps most evident in the different approaches taken with health-related inventions. Countries like those in Europe have concluded that methods of treating living bodies are so fundamental to human health that patents covering them are disallowed; countries like the United States have instead determined that there is a very real risk that such methods will be suppressed and concealed by inventors if they are not rewarded for disclosure, and so still accept the basic patent compromise. Some countries perform a similar accounting to conclude that laboratory-created organisms should not be patented, while others reach the opposite conclusion.

Is society better off by allowing Philip Leder to patent the oncomouse? Leder and his colleagues at Harvard University developed a biotech process in the early 1980s to create a species of mouse that was genetically engineered to be susceptible to cancers, hence its christening as the *"oncomouse."* The benefit to human beings is clear in that the oncomouse provides a biological model on which much valuable research has been conducted, leading to a better understanding of cancer. This understanding may have been achieved countless years or decades earlier than might have been needed without the oncomouse, resulting in untold numbers of human lives being saved from a horrific disease. At the same time, what many find to be horrifying is the assertion of monopolistic control over life-forms.

Irrespective of the specific rights that governments grant to inventors, the basic bargain remains at the heart of all patent systems: the invention is disclosed to the public in exchange for some government-backed right. The preamble to Brunelleschi's patent also highlights the basic nature of this right, which is a form of monopoly right that permits the inventor to prevent others from using the invention—as the preamble poetically explains, the fruit of the inventor's genius and skill may not be reaped by another without his will and consent.

The basic nature of this right is sometimes misunderstood. A common belief is that a patent assures the inventor of an unfettered right to produce, market, and sell the invention—in the parlance of patent law, the right to "practice the invention." This is, in fact, completely untrue. What patent rights grant is the ability to *prevent others* from engaging in those kinds of activities. As a result, many patent holders may be prevented from using even their own inventions without permission from other patent holders.

### ii.  Licensing Arrangements

This may be illustrated using a simplified example involving nanotechnology products. Without considering a variety of technical issues that might be raised concerning their patentability, suppose that when fullerene structures were first developed a patent application was filed and granted to Inventor A and covered all types of fullerenes. Later, nanotubes were developed by Inventor B and are broadly considered to be fullerene structures falling within the scope of coverage of that first patent. Because he holds the basic fullerene patent, Inventor A is entitled to prevent Inventor B from making and selling his fullerene nanotubes. But Inventor B is still entitled to file for her own patent on this improvement in technology and may be granted a patent that covers nanotubes. Indeed, the intent of the patent system is to publicize Inventor A's development of fullerenes so that others, like Inventor B, will have access to the information and be in a position to invent improvements on it.

This process may continue. After Inventor B's patent issues, yet another person, Inventor C, may dope the nanotube structures with certain materials that cause the structure to become superconducting. This represents still a further improvement in technology that was accelerated by the incentive for

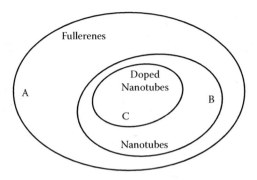

**FIGURE I.2**
The patents issued to A, B, and C are of progressively narrower scope. Because A's patent covers all fullerenes, B and C both need a license from A. Because B's patent covers all nanotubes, C additionally needs a license from B.

disclosure that the patent system provides. Inventor C may accordingly file for and be granted a patent that covers doped nanotubes.

The three patents that have now issued cover progressively narrower aspects of technology, but the narrower patents tend to cover technology that is more commercially useful. Consider the situations in which Inventors A, B, and C now find themselves. Inventor C, even though he holds a patent on the very valuable doped nanotube structures, may nonetheless be prevented from making and selling them because he infringes the patents held by Inventors A and B. He infringes the patent held by Inventor B because his doped nanotubes are still nanotube structures that are covered by the second patent. And he infringes the patent held by Inventor A because his doped nanotubes are also fullerenes that are covered by the first patent. For Inventor C to be able to make and sell these structures, he must obtain permission from both Inventors A and B, usually in the form of a license (see Figure I.2).

Inventor B faces a similar difficulty. Even though she holds a patent on nanotube structures, she infringes Inventor A's patent. Inventor B may be prevented by Inventor A from fabricating her nanotube structures because they are a type of fullerene covered by the first patent.

At first glance, it may appear that Inventor A is in the catbird seat. In many respects he is, for he owns the fundamental patent for a line of technology that has begun to develop and that has attracted public interest. Anyone who wishes to fabricate fullerenes will have to deal with him in some fashion, and he may be rewarded financially by selling licenses to his patent to Inventors B and C, and to anyone else who might want to participate in commercializing this line of technology. This is how it should be. Inventor A made a fundamental discovery that has spawned a whole new line of valuable technology. The patent system has put him in the position of being able to reap a significant financial reward in exchange for having disclosed his invention publicly and permitting others to improve on it.

Of course, the ability of Inventor A to exploit his invention is limited. The patent that he was granted is enforceable only for a limited time, after which anyone will be able to manufacture and sell fullerenes without his interference. Also, it may be that Inventor A recognizes the much better commercial value that exists for the discoveries of Inventors B and C. While the basic buckyballs that Inventor A initially discovered are viewed by the public as interesting curiosities, with their soccer-ball truncated-icosahedral shapes, consumers are really interested in getting their hands on nanotubes. The nanotubes that Inventor B developed have tremendous tensile strength that makes them useful in all sorts of industrial applications. The doped nanotubes that Inventor C developed are even more valuable and can command a higher sales cost because they not only have that tensile strength, but also have superconducting electrical properties that are in high demand.

Knowing that there is really little market for his buckyballs, Inventor A may wish to make and sell doped nanotubes—the demand for them is becoming extremely hot and, having read the disclosure provided in Inventor C's patent, Inventor A has thought of a few ideas to make them even more valuable. These are ideas that could, for instance, raise the temperature at which the nanotubes will be superconducting, allowing the use of specialized cooling units to be avoided.

But Inventor A now finds himself confronted with the same difficulties that faced Inventors B and C, namely that patents held by others may prevent him from marketing the doped nanotubes that he would like to develop. It is in such circumstances that cross-licensing arrangements may become attractive to all the parties involved. In these kinds of arrangements, Inventor A provides a license to his basic patent to Inventor C in exchange for a return license by Inventor C to the doped nanotube patent, and similar arrangements are made between A and B and between B and C. This allows all the parties to continue to develop and market products, with the public benefiting from the competition that the parties engage in to attract their business.

As time goes on, each of the inventors (and, of course, still other inventors) will build further on these ideas, with the patent system progressively encouraging disclosure of the new innovations to the public. Of course, the system is not a perfect system and many criticisms leveled at it have some merit. But while deficiencies with the system are relatively easy to identify, it is considerably more difficult to identify solutions that would not weaken the system by inhibiting inventors from disclosing their ideas. As it stands, the patent system represents a compromise that has been reached to try to balance competing considerations over a period of centuries. In considering the various tradeoffs—which at their core reflect the tension between the benefits of public disclosure with the detriments of monopolies—it is perhaps helpful to paraphrase Winston Churchill's comments about democracy as a system of government: "No one pretends that [it] is perfect or all-wise. Indeed it has been said that [it] is the worst form ... except for all those other forms that have been tried from time to time."

## Discussion

1. As mentioned, different countries permit or prohibit the patenting of methods for treating living bodies. What evidence is available to support each of these positions? Which position reflects the better public policy? Should nanotechnology inventions that are used in the treatment of human bodies be treated the same way as other methods of treatment, or are they different? Why?

2. While Brunelleschi's patent had a term of three years, modern patents typically have a term of twenty years. In some technology areas, such as software, it is sometimes suggested that this term is too long because of the rate of advancement of the technology. What is an appropriate term length for a nanotechnology patent? What difficulties do you see in having different term lengths for different technology areas?

3. Should the monopoly power granted to the patent holder permit her to prevent experimental uses of the technology by others in trying to improve the technology? How would such power defeat the goals of the patent system? Research the extent to which different countries exempt such uses from patent grants. Are such exemptions dependent on the type of technology? Should they be?

4. A patent cross-licensing arrangement permits a patent to be used defensively to avoid payment of a license fee. What factors might be relevant in quantitative valuations of patents obtained for such defensive purposes?

## 2. Patentability Requirements

Probably the most fundamental characteristic of nanotechnology is that it is concerned with the very small. Even the most cursory look at articles written about nanotechnology reveals just how pervasive the urge is to make puns about size. Headlines tell us about "The Small Wonders of Nanotechnology" and how "Nanotechnology is the Next Big Thing." The working title of this book was "A New Scale for Lady Justice," reflecting the author's own impulse to highlight just how central size is to any discussion of nanotechnology. As multiple other headlines tell us, when it comes to nanotechnology, "Size Matters."

How does this focus on size impact the ability to obtain a patent for a nanotechnology invention? For instance, in considering the general requirements for patentability discussed in the following section, it may be useful to keep a specific example in mind. In the autumn of 2005, nanotechnology researchers at Rice University reported that they had fabricated the world's first "nanocar."[2] This device represents a rather remarkable feat of nanoengineering. With four buckyballs acting as wheels, the nanocar has four independently rotating alkyne axles supported by an organic-molecule

**FIGURE I.3**
An illustration of a pair of nanocars. Each nanocar has four buckyball "wheels" connected to the "chassis" by an alkyne axle. (Image reproduced with permission of Rice University.)

chassis (Figure I.3). In the spring of 2006, a motor that responds to impacts by light photons was added to the car to provide a propulsion mechanism.[3]

Nanocars may have important practical applications. For instance, they may be used to carry cargo across nanoscale structures, providing a mechanism for precise movement and placement of materials at very small scales. Such approaches could improve fabrication of small structures, such as by permitting the materials needed in building up an electronic chip to be delivered by a fleet of nanocars more precisely than with conventional techniques and generating fewer defects than possible with traditional methods.

Part of the fascination with these nanocars is that they resemble macroscopic structures that we are all familiar with, and in some ways it is easy to understand the basic properties of their functionality. But at the same time, there were real technological barriers to be overcome in producing structures this small. And the precise mechanism by which they operate is very different from how the macroscopic counterparts operate: A nanocar has no internal combustion engine; it has no passenger compartment; it has no need for windshield wipers, air-conditioning systems, taillights, cupholders, or a host of other systems that conventional cars include. So, to what extent is a nanocar nothing more than a scaled-down version of something well known and to what extent does it represent something patentable? In what ways does patent law as it currently stands speak to these issues?

### i.   The Structure of a Patent

A patent has two principal components: a written description of the invention and a set of claims (Figure I.4a and b). The claims are intended to set

**(a)**

United States Patent [19]

Stephens

[11] Patent Number: 5,341,639

[45] Date of Patent: Aug. 30, 1994

[54] FULLERENE ROCKET FUELS

[75] Inventor: William D. Stephens, Huntsville, Ala.

[73] Assignee: The United States of America as represented by the Secretary of the Army, Washington, D.C.

[21] Appl. No.: 5,729

[22] Filed: Jan. 19, 1993

[51] Int. Cl.⁵ .................................. F02K 9/00
[52] U.S. Cl. ................................ 60/204; 60/208; 60/209; 60/253

[58] Field of Search ............. 60/200.1, 204, 208, 60/209, 253

[56] References Cited

U.S. PATENT DOCUMENTS

| H1234 | 10/1993 | Burgner | 60/253 |
| 4,133,173 | 1/1979 | Schadow | 60/204 |
| 4,332,631 | 6/1982 | Herty, III et al. | 60/208 |
| 4,355,663 | 10/1982 | Burkes, Jr. et al. | 60/253 |
| 4,574,586 | 3/1986 | Ostrych | 60/254 |
| 4,891,938 | 1/1990 | Nagy et al. | 60/264 |
| 5,133,183 | 7/1992 | Asaoka et al. | 60/204 |
| 5,152,136 | 10/1992 | Chew et al. | 60/251 |
| 5,234,475 | 8/1993 | Malhotra et al. | 44/282 |

| 5,239,820 | 8/1993 | Leifer et al. | 60/204 |
| 5,258,048 | 11/1993 | Wherwell | 44/282 |

OTHER PUBLICATIONS

Flood of Fullerene Discoveries Continue Unabated, Jan. 1, 1992, C&EN, pp. 25–33, Rudy M. Baum.

The Chemical Properties of Buchminsterfullerene (C$_{60}$) and the Birth and Infancy of Fulleroids, Fred Wudl, Acc. Chem. Res. 1992, 25, 157–161.

Fullerenes From Geological Environment, Peter R. Buseck et al., Science, vol. 257, pp. 215–216, Jul. 1992.

Primary Examiner—Richard A. Bertsch
Assistant Examiner—Howard R. Richman
Attorney, Agent, or Firm—Hugh P. Nicholson; Freddie M. Bush

[57] ABSTRACT

Solid fuel gas generator compositions for use in ducted rockets, in which fullerenes or substituted fullerenes are used. Fullerene compounds in which easily oxidizable groups, oxidizing groups, or salts of oxidizing acids are attached to the spherical carbon skeleton of the fullerene.

6 Claims, 1 Drawing Sheet

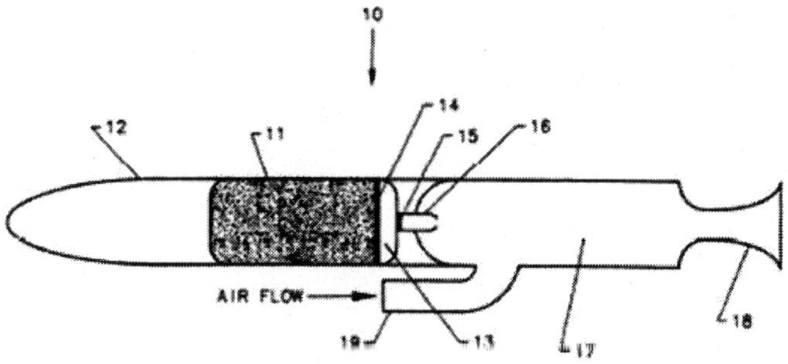

**(b)**

1. In the method of operation of a ducted rocket comprising a primary combustor containing a solid fuel gas generator for supplying hot fuel gases and partial decomposition products for combining with oxygen acquired from air during flight for producing complete combustion products in a secondary combustor and discharging combustion gases through a nozzle to provide thrust for said ducted rocket, the improvement in said method of operation of said ducted rocket for achieving high efficiency in the operation of said ducted rocket comprising incorporating into a solid fuel gas generator composition a solid fuel selected from the group consisting of fullerenes having a cage structure and derivatives of fullerenes having a cage structure.

**FIGURE I.4**

(a) A copy of the front page of U.S. Pat. No. 5,341,639, one of the first nanotechnology patents issued in the United States. (b) Claim 1 of U.S. Pat. No. 5,341,639.

forth in very specific terms precisely what it is that defines the invention. In this respect, they are the heart of the patent. Consistent with the view that patents are a form of intellectual "property," it is common to analogize the claims to a survey of real estate. The survey defines precisely where the boundaries of a particular property lie: If someone is within the boundary, she is a trespasser; if she is outside the boundary, she is not a trespasser. Similarly, the claims are intended to define the exact boundary of the invention: If someone develops a product that is within the boundary, he is an infringer; if he is outside the boundary, he does not infringe.

As always, such a simple rule is never quite as straightforward to apply in practice as it is in theory. For example, a real-property deed could identify one boundary by a natural feature such as a river. What happens if the course of the river changes over time? So too, with a patent, what happens if the meaning of a term in the claim changes over time? A deed could identify a boundary using a term that is ambiguous, setting the stage for an argument over the meaning of the term. In the same way, a patent claim may use terminology that different people construe differently, later precipitating arguments over exactly where the boundary of the intellectual-property rights lie. There are, indeed, countless ways in which arguments may be made over where the boundaries of the claims lie. Considerable effort is made to mitigate these possibilities by requiring that the claims be written as precisely as possible. This is reflected in a set of very inflexible rules that claims must conform with.

The written description of the invention is intended to provide support for the claims. It must describe what is being claimed in a manner that will place knowledge of the invention in the public domain. This is determined according to a number of criteria that are often treated distinctly, although they are related. In essence, the written-description portion of the application must provide enough detail about how to make and/or use what is being claimed such that it will be understood by those who are reasonably skilled in the technology. In the United States, this must include a disclosure of the "best mode" known to the inventors for practicing the invention, although many other countries do not impose this requirement.[4]

The combination of the claims and the written description must meet a number of specific requirements for a patent to be valid. The system that is used in most countries permits this validity to be considered at different points in time. The various requirements are considered initially by a patent examiner who reviews them when the patent is initially submitted to a patent office as an application. After the examiner concludes that all the requirements have been satisfied, the patent may be issued and enjoys a legal presumption that it is valid.

But this need not be the end of the story. That presumption can be attacked in a number of different ways. In some countries, third parties may bring new information to the attention of the patent office and ask that the patent office reconsider its determination. Perhaps more commonly, the validity of a patent may be attacked in a legal proceeding in court. In such cases, a judge or jury may be asked to review the patent in the light of evidence to determine whether all the criteria for patentability have actually been met.

Depending on how valuable the patent has become, this type of attack may be quite concerted. While the current cost for obtaining a patent is around $25,000, litigation of a patent to prove its invalidity may easily cost millions of dollars. Part of this cost may represent a near-literal scouring of the globe to prove that some requirement has not been met: Is there some obscure paper in a Chinese journal that describes a similar idea? Was some small start-up company in Greece that long ago went bankrupt making some similar product? Is the idea behind the invention described in a Russian doctoral thesis available only at a library in St. Petersburg? If an army of attorneys reads every document and e-mail found on every computer of a large corporation, can they uncover some proof that the inventor deliberately misled the patent office? The level of expense that can be justified in attacking a patent on any of a number of grounds is limited only by how valuable that patent has become in protecting the monopoly it grants in the marketplace.

### ii.  Patentable Subject Matter

So, what are the criteria that are considered by the patent examiner (and perhaps later considered by courts) in assessing a patent? First, the invention must be useful.[5] This requirement has a very low threshold, as evident from any of the many patents that occasionally elicit quiet chuckles. In the past, this requirement was sometimes used to reject inventions that were deemed "immoral"—gambling machines, certain types of weapons, sex toys, and so forth—but now, as long as there is a tiny modicum of utility, the requirement will be met (see Figure I.5).

In addition, the invention must fall within one of several specific classes of inventions: the invention must be a process, a machine, an article of manufacture, or a composition of matter. The Supreme Court of the United States has concluded that although this list is short, it covers a great deal. Indeed, it includes "anything under the sun that is made by man."[6] What it does not include are laws of nature, physical phenomena, and abstract ideas. Examples that the Supreme Court gave of "inventions" that were thus outside the list were a new mineral discovered in the earth, a new plant found in the wild, Einstein's mass energy equivalence law $E = mc^2$, and Newton's law of gravity.

In considering nanotechnology inventions, this requirement may thus sometimes come into play. A patent claim will be invalid if it tries to claim a nanotech structure that occurs naturally. Indeed, this requirement exposes one of the weaknesses of the example used in the previous section: Because fullerenes occur naturally in soot with other forms of carbon molecules, they could not properly be patented by themselves. Similarly, a researcher who

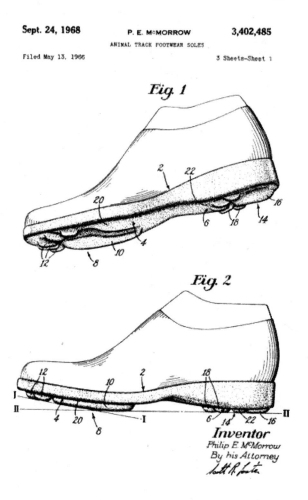

Sept. 24, 1968        P. E. McMORROW        3,402,485
ANIMAL TRACK FOOTWEAR SOLES

Filed May 13, 1966                          3 Sheets—Sheet 1

*Fig. 1*

*Fig. 2*

*Inventor*
*Philip E McMorrow*
*By his Attorney*

**FIGURE I.5**
A copy of the front page of U.S. Pat. No. 3,402,485, issued on May 13, 1966. The animal track footwear soles described in this patent allow a wearer to lay simulated animal tracks. This is one of many illustrations of how small a level of utility is required to meet that part of the requirement for patentability.

discovers a new aspect of quantum mechanics that opens the door to all kinds of exciting nanotechnology developments would not be able to patent it if it represents a law of nature.

Sometimes, this subject-matter requirement can be finessed. While it may not be possible to patent fullerenes themselves, an efficient method for producing fullerenes may legitimately be patented as a *"process."* This is not merely a semantic distinction. Purely natural processes may be horribly inefficient, producing fullerene structures rather randomly with all kinds of other molecules. The development and refinement of a process that is more selective in the types of molecules produced is a result of the ingenuity of human beings.

### *iii.* **Prior Art**

In addition to the utility and subject-matter requirements, inventions must be new[7] and cannot be obvious[8] over what was previously known. These requirements represent by far the major focus in evaluating the validity of patents, both during examination in a patent office and during litigation. The body of previous knowledge against which novelty and obviousness are measured is referred to as the *"prior art."* Precisely what information is included in the prior art reflects a number of different policy determinations. While the following discussion emphasizes those policies that have been adopted in the United States, other countries do emphasize different concerns so that the scope of the prior art sometimes varies from country to country.

The different sources of prior art may be considered in three broad groupings: documentary references, U.S. patents, and everything else. The first group, documentary references, essentially includes everything that has been published in written form, irrespective of where the publication took place. It includes journal articles, newspaper articles, patents that have issued (or patent applications that have been published) in foreign countries, everything that has appeared on the World Wide Web, and many other sources of information. For a patent claim to be valid, it cannot have been disclosed in any of these sources nor be obvious over what has been disclosed in these sources. There are wide differences in how accessible these different documentary sources are—material published on a popular Web site will be much more widely known than information published in an obscure technical journal in a language understood by only a handful of people—but the patent laws treat them entirely equally.

In most countries, a patent cannot be obtained if the invention was disclosed in a documentary reference before the patent application was filed. The United States is unusual in that it affords a grace period between the publication of the documentary reference and the filing date of the application. This grace period is one year: as long as the material was published less than a year before the filing of the patent application, a documentary reference will not compromise the ability to obtain a patent in the United States. This provides inventors with the freedom to disseminate their ideas for at least a limited time without relinquishing their patent rights. A handful of other countries also provide a grace period, usually somewhere between six months and a year, but the vast majority of countries provide no grace period at all. In these countries, if no application has been filed, patent rights are forever lost the day after any kind of publication of the invention.

Special status is given to documentary references that are in the form of U.S. patents (or publications of U.S. patent applications). They accordingly form the second group of prior art. This special status is based on a fiction that the U.S. Patent Office should operate with perfect efficiency. The idea is that once a patent application is filed, the patent office should immediately be able to determine whether to grant the patent, and do so on the very day the application is filed. Any delays in actually doing so are ignored—U.S. patents and published U.S. patent applications are therefore treated as prior art as of their filing date.

The third group of prior art includes everything else: actual public uses of the invention, sales of the invention by someone, oral disclosures of the invention at a conference, and so forth—it includes essentially anything unwritten that resulted in the invention being known in a public way. In the United States, this third group is considered to be prior art only if the unwritten disclosure occurred within the territorial borders of the United States. This is a somewhat parochial position to take and reflects a policy judgment that permits the patenting of inventions that, although known somewhere in the world, are known only in the most parochial of fashions.

That is, if some invention was being used publicly in Europe but news of it had not yet reached the United States, it would remain patentable in the United States as long as it was not disclosed in a public documentary reference. This is true irrespective of how long or how common the invention was in Europe—if it was known there only through unwritten sources and was unknown in the United States, it remains patentable here. This would similarly be true if actual devices were being manufactured and sold routinely in Australia (or anywhere else in the world)—as long as knowledge of them is limited to unwritten sources and they remain unknown in the United States, they remain patentable.

Another policy judgment that the United States makes is that information in the second and third groups—that is, everything other than documentary references—is prior art only if the disclosure occurred before the date of the invention, irrespective of when the application describing and claiming the invention was filed with the patent office. This policy embraces the idea in the United States that a patent for an invention should be awarded to the first person who actually invents it, rather than to the person who wins a race to the patent office. When an inventor is confronted with potential prior art from the second and third groups, a mechanism therefore exists for him to prove that the time of his actual invention predates those disclosures.[9]

The determination of what is and what is not prior art is not always simple (Figure I.6). Some disclosures have aspects that fall into more than one of the three groups outlined above, and careful consideration may be needed to reconcile the different policy objectives. In addition, I have simplified certain complexities that exist in the law to avoid obscuring the fact that the specific rules directly embrace certain policy considerations that the United States finds important—that documentary references are more important than other types of information; that when there are competing claims to an invention, the first to make the invention is the one who should be awarded the patent; that unwritten knowledge should become relevant only when it penetrates the borders of the United States; and so on.

The calculus by which the United States reaches these policy conclusions is different from how many other countries evaluate similar issues. The result is that the same disclosure might be prior art in one country but not in another.

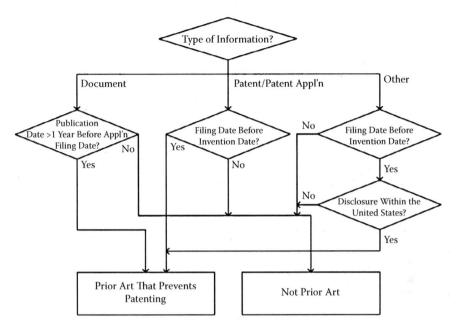

**FIGURE I.6**
A flowchart summarizing how to determine whether information is prior art.

In turn, this has the consequence that some inventions may be patentable in some countries even while not being patentable in other parts of the world.

However the prior art is ultimately defined in any given country, the twin requirements of novelty and nonobviousness[10] must be met for an invention to be patentable. Whether an invention is novel is by far the easier of these requirements to apply. The concept of obviousness is much more difficult. Something that is obvious to one person may well not be obvious to another, particularly if these people have different levels of expertise in the technology. But most insidious is that determining whether some advance was obvious is always done after the fact. We all have experiences where we were initially unable to grasp some concept or to see the solution to some problem. But after it is explained to us, we suddenly find ourselves in a state of astonishment that we were unable to see the answer because it now seems so simple and apparent.

This natural tendency to view circumstances differently with the benefit of hindsight is very much recognized by courts. Much effort has therefore been expended to set forth an objective standard by which obviousness can be evaluated. The essence of this standard attempts to place us back in the shoes of the inventor by focusing on what was known in the prior art. An invention is obvious if all its elements were known in the prior art and there was some reason to combine them in the manner of the invention that can be reasonably articulated.[11] Although there is no requirement that it be applied rigidly, one recognized way of articulating such a reason is to show that there was some teaching, motivation, or suggestion in the prior art itself to make the combination. In this way, the standard recognizes that many inventions result from combining elements that are well known. In considering the creative process of inventing, one court has aptly observed that "[o]nly God works from nothing. Men must work with old elements."[12]

### iv. Obviousness and Scale Factors

An issue of particular relevance to nanotechnology is the extent to which the law considers it obvious to reproduce something that is known, but at a different scale. There are some significant indications that merely changing the scale of some known structure or process is not, by itself, patentable.[13] This is a reasonable position in many circumstances. For example, if changing the scale involves no new technological issues and provides nothing that can in any way be characterized as unexpected, it hardly seems to warrant a patent.

Consider, for instance, a toy manufacturer who builds an exact replica of a Ferrari, but so that it is roughly the size of a large dog or pony. Such a mini-Ferrari might find some success in being marketed as an actual working car. In this example, the Ferrari replica is identical in every way to a full-size Ferrari—it has exactly the same structure, right down to the least significant nut and bolt. This replica might represent a spectacular feat of engineering to reproduce the structure so exactly at a smaller scale. But if the only challenge in making it was scaling down something that was already known, it does not represent a patentable invention.

Now contrast this circumstance with development of the nanocar described at the beginning of this section. The nanocar is hardly a replica of a macroscopic-scale car—it does not function in anything like the manner of a conventional automobile and has a significantly different structure. Indeed, the reasons why it is described as a nano*car* are largely superficial and mostly represent an attempt to draw a cute analogy with technology that is widely familiar. The development of the nanocar is patentable over the prior-art macroscopic car because the technical challenges that needed to be overcome were not expected from what was known about conventional cars. It was not simply a matter of taking the blueprints for a Ferrari and scaling them down. Specific challenges were presented in the need to accommodate a variety of small-scale effects that are inapplicable at large scales—the effects of molecular interactions on the scale of the nanocar itself, quantum mechanical effects, an entirely different form of "friction," and a host of similar issues.

What is relevant in evaluating the patentability of the nanocar is whether there was anything unexpected: Were there manufacturing difficulties that were unforeseen in light of prior art cars? Were there advantages in ways of using the nanocar that could not have been predicted from the structure of prior-art cars? Were the ways in which the nanocar could be assembled surprising, even to experts in assembling conventional cars? It is an affirmative answer to these types of questions that points to the nanocar not being obvious over the prior art, and therefore patentable.

Of course, a comparison between conventional cars and nanocars is a rather extreme example. A nanocar is roughly one-billionth the size of a conventional car, making it easy to see that there would be a number of unexpected aspects to achieving its construction. In reality, the comparison between a nanotechnology invention and the prior art is never going to involve such a large difference in scale. The area of nanotechnology that is most likely to confront issues of scale changes is that of nanoelectromechanical systems (NEMS). These are mechanical structures made on the nanometer scale that are driven by electrical energy—very small gears, pumps, sensors, actuators, and so forth.

The way such structures are constructed may be very similar to the way microelectromechanical systems (MEMS) are constructed, using the same basic techniques and similar types of component parts. In comparing NEMS with MEMS, the difference in scale might be only a factor of ten or so—comparable to the scale difference between a Ferrari and the replica described earlier; the physical issues that are confronted could therefore be very similar. Indeed, researchers in the area of small electromechanical systems sometimes have difficulty in characterizing a particular structure as a MEMS device versus a NEMS device because there is a continuum in scale with no clean division between the two. Attempts to patent NEMS devices could very well confront difficulties posed by relevant prior-art MEMS devices. The same calculus described earlier applies—if the NEMS device (Figure I.7a) is just a scaled-down version of a MEMS device (Figure I.7b), was something unexpected encountered? It is easier to see that the answer to this question will be "no" when the scales of the devices being compared are closer.

## v. Recap

For a patent application to be allowed, it must meet every one of the criteria outlined earlier. Each and every claim in the application must define an invention that represents patentable subject matter, is useful, is novel and not obvious over the prior art, and is supported by a written enabling description that sets forth the best mode for practicing the invention. Any claim that fails to meet each of these requirements is invalid.

The role of the patent examiner is to review the application and the prior art to ensure compliance with these requirements. This examination has an important role to play because it affords claims in issued patents a legal presumption that they are valid. The very fact that the claims have been reviewed by an examiner and found to meet every one of the requirements described above increases the burden on someone later trying to assert that a claim is not valid. If there is a close call trying to decide whether a claim is valid, this presumption will cause the decision to fall in favor of the patent holder.

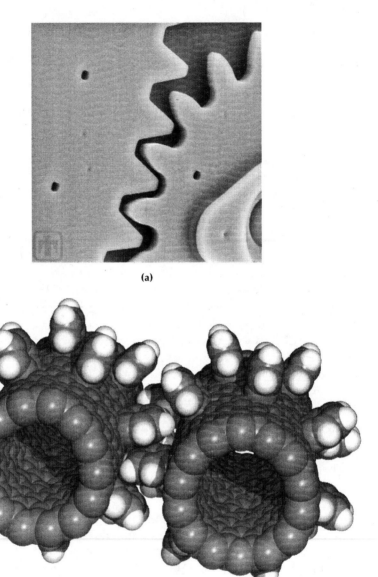

**(a)**

**(b)**

**FIGURE I.7**
(a) A gear constructed at MEMS scales. (b) A drawing of a gear constructed at NEMS scales.

## Discussion

1. Sometimes two people independently invent the same thing. In most countries, the first of these people to file a patent application with the patent office is the one entitled to the patent. The United States is

alone in taking a different view and grants the patent to the first one who actually completed the invention. Does one approach disadvantage small inventors? Which approach better meets the public-policy objectives of the patent system?

2. In the United States, patent applicants are required to disclose to the patent office any information that may be material to examination of an application. Most often this takes the form of providing prior-art references to the examiner that the applicant is aware of and that might bear on patentability issues. In most other countries this duty does not exist, but members of the public are given an opportunity to oppose the grant of a patent before it issues by presenting evidence of unpatentability. Which system is more likely to prevent the allowance of invalid claims? Discuss ways in which each of these systems might be subject to abuse. What types of reforms would reduce the ability to engage in those abusive practices?

3. Figure I.7 compares a MEMS gear and a NEMS gear. What arguments can you develop that the NEMS gear is patentable over the MEMS gear? Do you think those arguments would be effective? What additional evidence could be submitted to overcome the weaknesses in your arguments?

## 3. Riding the Patent Office Pony

In the early days of the United States, the procedure for determining whether to grant a patent to an inventor was remarkably streamlined. From a modern perspective, it seems almost quaint. Signed into law by the president on April 10, 1790, the first U.S. patent statute directed applicants to file a petition with the secretary of state, who would then vote with the secretary of war and attorney general whether they thought "the invention or discovery sufficiently useful and important."[14] If a majority of them thought so, a patent would be awarded to the applicant. To be sure, there were requirements that the application had to meet in the form of a written description, drawings, and possibly a working model. And this committee quickly began to generate standards by which to evaluate applications. But the overall procedure has the air of having been rather informal.

### i. The Current State of the U.S. Patent Office

This seems especially to be the case when considering the current state of the U.S. Patent and Trademark Office, which employs on the order of 5000 examiners. This army of examiners is responsible for the examination of the hundreds of thousands of applications that are filed every year. In 2006, the patent office received more than 440,000 new applications for patents but completed examination of only 332,000 applications—a shortfall of some 100,000 applications that contributed to the existing backlog.[15] Applications are filed by applicants from around the world in every conceivable technology area. In some ways, this is a testament to the success of the U.S. patent system. The widespread desire to obtain U.S. patents highlights the significant value that they are seen to have, not only within the borders of the United States, but throughout the world. At the same time, the volume of applications that must be considered by the patent office is ever more frequently described with language reminiscent of disasters having biblical proportions.

Nanotechnology applications have been part of the increase, and even out-pace the overall rate of increase to the patent office as a whole. Figure I.8 provides an illustration of the growth in nanotechnology applications that have been submitted to the U.S. Patent and Trademark Office over the last fifteen years. The trends seem to evidence having passed a barrier in which commercialization of nanotechnology inventions is now possible.

The data used to generate the graph in Figure I.8 was collected by searching issued patents and published patent applications for the appearance of certain terms that are related to nanotechnology. One reason that this was done is that, until very recently, the patent office provided no specific classification for nanotechnology. The reason for this is at least twofold—first, the technology is so new that a particular need for a specific nanotechnology classification is only beginning to be recognized; and second, nanotechnology is, by its very nature, highly interdisciplinary. At the same time that this interdisciplinary nature has real benefits—permitting the sharing of relevant information by those from different areas of expertise—it also resulted in a certain fragmentation of how nanotechnology patents were evaluated by the patent office.

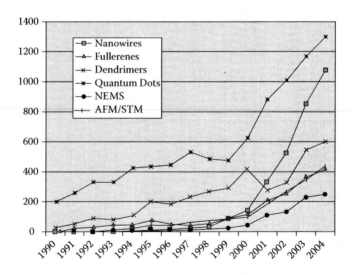

**FIGURE I.8**
A graph of nanotechnology filings to the U.S. Patent and Trademark Office. Data before 2001 collected from issued patents; data after 2001 collected from published applications.

## ii.  Classification of Nanotechnology Patent Applications

Because the patent office is so large and handles so many applications, the earliest nanotechnology patents were considered by those parts of the office that traditionally handled the technology from which the applications grew. A nanotechnology application on quantum dots developed by a semiconductor physicist may have been considered by an examiner in a group that handles semiconductor structures; an application directed to biological applications of dendrimers may have been considered by an examiner in the biological arts; an application directed to nanophotonics may have been considered by an examiner who routinely works with optics; and so on. This fragmentation has the potential drawback that an examiner will be unaware of relevant information that has been developed in another field and that the examination of the patent application will suffer for it. A current distribution of nanotechnology-related patent applications in the United States shows that about 30 percent are classified as "electrical" inventions, 26 percent are classified as "biotechnology" inventions, 23 percent are classified as "chemistry" inventions, and 21 percent are classified as "mechanical" inventions.[16]

The patent office has engaged in a number of efforts to address this fragmentation. For example, in November 2001, the patent office initiated a project to identify nanotechnology-related inventions and used this to develop a cross-reference classification digest in August 2004. This digest was intended to simplify the searching of documents relevant to nanotechnology, particularly by patent examiners in evaluating the novelty of newly submitted patent applications. This was superseded in November 2005 by the establishment of a more complete cross-reference that includes 263 subclasses to permit more refined classification and searching.[17] This is likely still an intermediate step to development of a complete classification schedule that is as detailed as the classification schedules that the patent office has in other technology areas.

Input into the development of a suitable classification system was one result of a series of Nanotechnology Customer Partnership meetings that the patent office initiated in September 2003. The meetings are informal gatherings among potential nanotechnology patent applicants and representatives of the patent office. They have been held every few months and provide a mechanism for the patent office to receive feedback that it uses in its attempts to improve the overall examination of nanotechnology applications. It also provides a venue for the patent office to articulate where it currently stands in addressing issues of particular concern to nanotechnology applicants—how examiners will be identifying relevant prior art, how issues like changes in scale will be addressed, and so on.

Every one of the thousands of applications filed with the patent office every day has the potential to mature into a patent and therefore to provide a government-backed monopoly power that could have wide-reaching commercial implications. Each application has the potential to affect the cost of goods and services within the United States, to impact the success or failure of commercial entities (and therefore the employment of the people who rely on those entities for their livelihoods), and to inhibit or promote future lines of scientific and technological development in the country. While the ability of different applications to realize this potential will of course vary, each submission deserves to be treated with these potential ramifications in mind. It is therefore important that each application be considered with care to determine whether it meets the criteria for patentability—significant adverse consequences can result both from issuing a patent that is not valid and from refusing to issue a patent that complies with all requirements.

### iii.   Security Review and Secrecy Orders

The patent office is highly compartmentalized. When an application is initially submitted to the patent office, it is considered by the Office of Initial Patent Examination. This office is responsible for checking for compliance with formalities, preparing the application for entry into various databases used by the patent office, and for assigning the application to the correct technology center for examination. In addition to these functions, the content of the application is reviewed at this stage to determine whether it includes any subject matter that various government agencies have identified as being of interest.

If so, the application will be made available to those agencies who will review it for the possible imposition of a secrecy order.[18] Patent applications are on the forefront of technological change and orders for secrecy may be desired by the government for any of a variety of reasons. Nanotechnology applications are particularly likely candidates for secrecy, especially those that describe inventions that may have military uses, intelligence-gathering uses, and so on. If a secrecy order is imposed, there may be a number of ramifications. Perhaps most obvious is the fact that the information in the application will be prohibited from being disclosed publicly—the application will not be published and the applicant will not be permitted to disclose the information through other means. This includes a prohibition on filing a counterpart to the application in other countries whose governments would then also have access to the information.

This secrecy also means that a patent cannot be issued on the application until the secrecy order is lifted. The application may undergo examination, but will be held up by the secrecy order even after it has been determined that it is patentable. The examination itself will be largely similar to conventional examination, except that it will be conducted by an examiner who holds a security clearance at an appropriate level.

It is well recognized that the effects of a secrecy order can be severe on an applicant. A company might be structured completely around the invention that is disclosed and will be prevented from engaging in its usual activities to commercialize the product. Applicants whose patent applications are suppressed by a secrecy order are entitled to compensation for financial damage that results from the secrecy order or by use of the invention by the government.[19]

### iv.  Examination

Irrespective of whether a secrecy order is imposed, the patent application is assigned to an examiner who is charged with evaluating whether the application meets the criteria for issuing a patent that were described in the preceding section. The particular specialization of the examiner is generally quite narrow—an examiner could be assigned to review only applications related to the manipulation of quantum-dot structures using atomic force microscopy or something similarly narrow. The advantage of such specialization is that the examiner becomes very familiar with both the applications that are being filed in his area and the prior art that is relevant to their examination.

The examination of the application usually takes the form of some back-and-forth between the examiner and the applicant. While it does sometimes happen, it is relatively rare that an application is immediately allowed. Instead, the examiner finds some reference that raises a question as to the novelty or obviousness of at least some of the claims. The applicant then explains why the reference is not as relevant as the examiner believed or alters the claims in some way to provide distinctions that will meet the patentability requirements. This back-and-forth can proceed a number of times, with one of three potential results: the examiner is persuaded that the application, perhaps with some amendment, meets the requirements for patentability; the applicant concedes that the examiner is correct in assessing the invention as not being patentable and decides not to pursue the application; or there is a stalemate with neither the examiner nor the applicant being convinced of the other's position.

There are numerous mechanisms to permit the examiner and the applicant to understand each other's point of view. While most of the interaction takes place in writing, opportunities are available for other types of interaction—interviews between the examiner and the applicant (or, more likely, the applicant's attorney) may be conducted by telephone, through video conferencing, or in person at the patent office. This type of interaction may be tremendously valuable, particularly when an applicant travels to Washington DC to take the time to explain face-to-face with the examiner what the invention is and how it differs from what was done before.

Without such direct interaction with at least some applicants, there would be a great risk that patent examiners would treat applications too abstractly— they would operate in a sheltered environment where they do not realize how strong an impact their decisions can have, where the ability to obtain a patent might determine the fate of a small company and its ability to maintain a certain level of employment. Even with the possibility of direct interaction such a risk remains. But a practical approach to interacting with the patent office recognizes that particular applications can be distinguished in the examiner's mind—by talking to the examiner on the telephone about the application or by visiting her in her office with the chief technology officer of the company that needs the patent to raise further investment capital.

There are other practical considerations that good patent attorneys make in helping applicants interact with the patent office. These attorneys are keenly aware of the environment in which patent examiners work; indeed, some of the most effective patent attorneys were once examiners themselves and know that environment intimately. The productivity requirements imposed on examiners are significant. For each application they consider, they are allotted a total time of about twenty-four hours to do the full examination—to read the application and learn what the invention is; to conduct a search of the prior art in the context of each claim; to determine whether each of those claims is novel and not obvious over the results of that search; to evaluate the formal requirements of the claims and of the written description; to prepare potentially multiple written communications to the applicant detailing the results of these analyses; to discuss the reasoning with the applicant over the phone or in person; and a host of other functions that are involved in examining a patent application.

This time allotment is quantified in a point system in which points (referred to in the patent office as *"counts"*) are accumulated for performing specified functions that typify a certain stage in the consideration of an application (and having spent the time to get to that stage). The performance evaluations of examiners, and the availability of salary increases and promotions, are directly tied to this measure of productivity. Good patent attorneys are aware of the motivational effect this system has on examiners and exploit it. Examiners have considerable discretion to be clear or vague in suggesting approaches that will address the concerns that exist, or to be generous or stingy in how they interpret claims and prior art. Establishment of an implicit quid pro quo during the process may result by taking actions that cause the examiner to earn a count.

Is exploiting the natural human motivations that such a point system creates improper? Almost no patent attorney would think so. The role of the patent attorney is to play the game as effectively as possible to secure the best possible rights for his clients. After all, it is the government that gets to set the rules of the game, and it can change the rules if it wants to.

If a stalemate has been reached between an applicant and the examiner, it is possible to go over the examiner's head by appealing the rejection to the Board of Patent Appeals and Interferences. This is a body made up of a group of specialized administrative law judges called patent law judges. They are highly experienced in applying the law to patent applications and have their own technical specializations. When the board hears an appeal, it does so as a panel drawn from the full body, usually with three of the judges considering a particular application (although sometimes more judges may participate if the appeal raises unusually important issues).

### v.  Appeals

In considering an appeal, the selected panel has access to the entire written exchange that has been developed between the applicant and examiner. This is supplemented with an appeal brief, which summarizes the applicant's views, and an examiner's answer to the appeal brief, which summarizes the examiner's views. These documents are especially valuable to the panel because there may have been some evolution in the positions of the applicant and examiner over time. Agreement may have been reached on some issues; some concerns may have vanished because of ways in which the claims were amended; and the specific character of certain issues may have changed for any of a variety of reasons.

The appeal brief and examiner's answer act to focus the attention of the panel on the precise issues that need resolution. In most cases, the applicant may also file a reply brief that responds to the examiner's answer, perhaps giving the applicant the advantage by getting in the last word. It is also possible to conduct an oral hearing before the panel. This may be especially valuable when the issues being raised are close ones since it provides the panel with the opportunity to present their questions to the applicant and the examiner. The way in which those questions are answered may well alleviate concerns that the individual judges might otherwise have in deciding close issues.

Not surprisingly, the board dislikes spending its valuable time considering arguments that are frivolous or unreasonable. The board exercises greater control on the type of arguments that may originate from examiners than it does from applicants. This is done by requiring that the examiner hold a conference with two other senior examiners before the appeal is transmitted to the board. This conference must result in agreement with the position taken by the examiner, with the two other participants also signing the examiner's answer.[20]

The board has a number of options it may take, the most straightforward of which are simply to uphold the rejections or to reverse them. Other actions that may be taken include a hybrid approach in which the rejections of some claims are upheld while others are reversed, or to request clarification of the arguments made by the examiner and/or applicant before reaching a decision.

In the event the board upholds the rejections, the applicant still has recourse to further appeals, but these are presented to the court system. At this stage, a rejection may be appealed to the Court of Appeals for the Federal Circuit, a court that specializes in patent cases.[21] If the appeal is unsuccessful in the federal circuit, a further appeal may be considered by the United States Supreme Court, although that court generally has the ability to be selective about which cases it is willing to hear. An alternative route to appeal a rejection in the court system is provided by the ability to bring a civil action against the director of the patent office in the United States District Court for the District of Columbia. Subsequent appeals in the event of an adverse decision there may then subsequently be had in the Court of Appeals for the Federal Circuit and potentially in the United States Supreme Court.

It is apparent that there are many steps that must be taken by the patent office in deciding whether to grant a patent (Figure I.9). Each of these steps inherently takes significant time, resulting in a long wait by patent applicants for a decision to be made. Currently, this wait is exacerbated by the general burdens placed on the patent office from the sheer volume of applications being submitted. A realistic time to obtain a patent from the time an

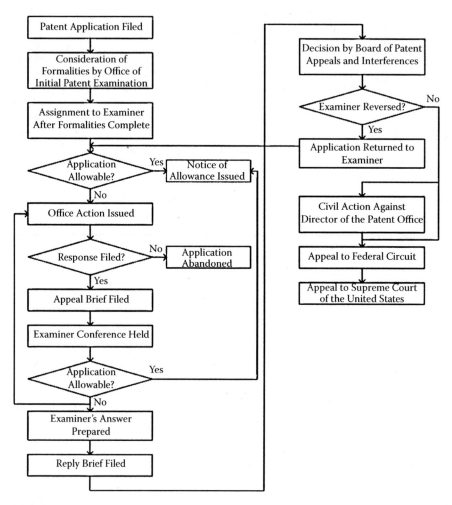

**FIGURE I.9**
Consideration of a patent application.

application is initially filed is somewhere around three to six years. In a fast-moving technology area like nanotechnology, this means that patents may be issuing on technology that is by then very much historical.

### vi. "Special" Applications

At the same time, nanotechnology applications may be the beneficiaries of a program that affords certain applications *"special"* status.[22] Once designated as "special," an application jumps the general queue at all stages in consideration by the patent office so that it is considered out of turn. This may greatly shorten the length of time to complete the consideration of the application. While nanotechnology itself is not currently a basis for affording an application special status, the diverse breadth of nanotechnology lends many of its activities to areas that may be granted special status. If expedited consideration is a concern, it may be desirable to prepare an application so that it embraces one of these areas.

For example, applications that describe inventions that "materially enhance the quality of the environment of mankind" may be afforded special status. The potential for nanotechnology to have a positive effect on the environment is often heralded. For example, manufacturing techniques at the atomic level are so precise that they generate considerably less waste than other types of techniques. In more proactive applications, nanotechnology sensors may improve the ability to monitor environmental impacts and to provide better controls on the use of materials and the generation of waste. Nanotechnology treatment systems may be directed at breaking down toxic substances. All of these applications may be characterized in a way that allows them to be afforded special status by the patent office.

A related class of applications that may be afforded special status are those that either contribute to the discovery of new energy resources or contribute to more efficient use and conservation of existing resources. The second of these categories may be particularly relevant to nanotechnology inventions that provide new types of solar or fuel cells, that provide energy savings through the development of lightweight composites, or that provide new and efficient kinds of energy-storage devices.

Inventions directed to superconducting nanotube structures are obvious candidates for special status as "involving superconductivity materials." Nanotubes doped with alkali metals are well known to be superconducting, and more recent research into multiwalled nanotubes suggests that even pure-carbon nanotubes may be engineered to be superconducting. Current efforts in the development of *"spintronics,"* which attempts to marry superconductivity and nanotechnology in developing systems that carry information with electron spin instead of with electron charge, may involve inventions within this category. It currently appears that the future potential for spintronics applications is limited only by the imaginations of researchers, with spintronic counterparts to virtually every electronic component (transistors, diodes, filters, etc.) having already been made.

A host of nanotechnology inventions may find application in counterterrorism, which is another class of inventions that warrants special treatment. Examples of such inventions include a variety of sensor-network deployments that provide information on the status of railway systems, roadways, bridges, water-distribution systems, cargo containers, and so forth. Not only can such sensors be developed to act passively in detecting chemical, biological, or radioactive materials, they can function actively in detecting movements in ports, nuclear power facilities, and other sensitive locations. Special fabrics that result from nanotechnology developments may be adopted by law enforcement or military personnel responding to terrorist activity;

research is actively being conducted into fabrics that have decreased receptivity to chemical and biological agents and resist penetration by projectiles.

There are also numerous biological applications being investigated for nanotechnology, including such devices that may provide very early cancer detection by being specially engineered to bind to molecules associated with cancer. Other devices may be used to deliver carefully controlled doses of chemotherapy agents directly to the cancer cells to be destroyed, relieving cancer patients of the need to undergo the widespread poisoning of their bodies associated with conventional chemotherapy. In other medical areas, there have been suggestions that fullerenes may be used to interfere with the ability of human immunodeficiency virus (HIV) to reproduce, offering a potential treatment for AIDS. These kinds of inventions may be entitled to special treatment by the patent office because they are related to HIV/AIDS or to cancer.

The interface between nanotechnology and biotechnology may also result in nanotechnology inventions being afforded special status because special provision is made for certain biotechnology inventions.[23] *"Biotechnology"* is not specifically defined in this provision so that a broad interpretation would permit virtually any nanotechnology invention that has a biological application to apply.

There are numerous ways in which nanotechnology applications may receive different treatment by the patent office , ranging from the manner in which they are classified and the way in which secrecy orders may be imposed to their official designation as special applications.[24] At the same time, there are many ways in which the consideration of nanotechnology applications by the patent office is entirely routine since the same basic process of examination is applied to them. But all patent applications are now considered by a large, sophisticated—and often bureaucratic—office.

The difference between the current patent office and the tribunal that voted on patentability in the earliest days of the U.S. patent system is as profound as the difference between nanotechnology and the types of inventions being considered at that time. Today the U.S. Patent Office issues some thousands of patents on Tuesday of every week. Long gone are the picturesque days when the State Department of the United States fed and housed a single pony that would be ridden by a messenger to receive the signatures needed to issue a patent. But the metaphor that arose from that practice, of "riding the Patent Office pony" as being synonymous with working there, still persists in some circles.[25]

### Discussion

1. Today, most industrialized countries use an *"examination"* patent system in which applications are examined before patents are issued. While such patents may still be challenged in court, they are issued with a presumption that they are valid; we noted in the previous section that this presumption may present certain difficulties in

challenging the patents. Many countries use a *"registration"* patent system in which no examination takes place at all. Courts provide the only venue in which validity is considered. The United States used such a registration system between 1793 and 1836. Compare the merits and disadvantages of examination and registration systems. Does the large volume of patent applications currently being filed argue in favor of the United States switching back to a registration system or in favor of retaining an examination system? What would be the effect on the court system if a registration system were adopted? What would be the reaction of business owners?

2. In determining whether to publish papers in scientific journals, editors almost universally rely on a peer-review process. In contrast, examination of a patent application is performed solely by the patent examiner without the benefit of peer input. Identify potential benefits and drawbacks of adopting a peer-review system for patent examination. You may wish to consider such issues as whether the scale of submissions to journals versus patent offices warrants a different treatment; whether there are differences in confidentiality concerns between the two systems; whether the potential for abuse by referees (and the consequences of abuse) differ between the two systems; and the extent to which scientifically accomplished but legally untrained referees can offer meaningful input to patent-examination decisions. Would the value of peer review of nanotechnology patent applications be different than, say, of applications for improvements in luggage or furniture design or applications for new types of cat toys?

3. The text suggests that because it is the government that sets procedures in the patent office, and has the power to change them, applicants and attorneys can ethically do as much as they wish to advance their interests as long as they stay within the confines of those procedures. Is this a fair assessment? Won't there always be loopholes that can be exploited, no matter what system of procedures the government develops? Should applicants and attorneys have their own responsibility to ensure that the process is consistent with the goals of the patent system? How would you go about policing such a requirement? Would you modify the way in which patent examiners are evaluated? How?

4. The ability of patent offices to examine nanotechnology applications is highly dependent on their effectiveness in recruiting and retaining examiners with sufficient relevant technical capabilities. What global economic factors are relevant to how this issue is manifested for different countries? What steps can governments take in resolving the challenges that face different patent offices in recruitment and retention? How significant is the fact that in most countries a patent examiner must be a citizen of that country? You may wish

to consult the book in this series by Louis Hornyak that discusses educational and workforce issues as they relate to nanotechnology for relevant data.

5. The discussion of the process for patent applications discussed in the main text applies to what is sometimes referred to as a *"non-provisional"* patent application. This term arises because the United States also allows the filing of *"provisional"* applications. Such provisional applications have no formalities associated with them and are not examined. But if a nonprovisional application is filed within a year of the provisional application and makes reference to it, the nonprovisional application may effectively enjoy the filing date of the provisional application for whatever was disclosed in the provisional application. Why do you think such applications are permitted? What benefits do you see to filing a provisional application? What risks do you see?

## 4.  Infringement Issues

There is an old Calvin and Hobbes comic strip in which Calvin hears music at Christmastime proclaiming that "He sees you when you're sleeping, he knows when you're awake. He knows if you've been bad or good, so be good for goodness sake!" This causes Calvin to pause and wonder about the true nature of Santa Claus: "Kindly old elf, or CIA spook?" These days, a similar question might be asked about Thomas Alva Edison: Wonderfully gifted inventor or repulsive patent troll?

### i.  Patent Trolls

Coining of the term *"patent troll"* is usually attributed to Peter Detkin, who was associate general counsel of Intel at the time, and describes an entity that relies on patents as intellectual property to extract licensing revenue without even making its own products. The term was coined as a substitute for *"patent extortionist"* after Intel had been sued for libel in using that term to describe two companies that had filed suit against it for patent infringement.[26] The implication of the term *"patent troll"* is that the practice it describes, while always acknowledged to be perfectly legal, is somehow unethical. The underlying assumption is that those who own patents should at least be involved in the commercialization of the products that they cover.

No one would seriously denigrate the inventive contributions of Edison by describing him as a patent troll. But it is true that Edison rarely commercialized the products described in any of his 1093 patents and that he engaged in behavior that is included in the broadest definitions of a patent troll. What is also undoubtedly true is that those attempting to commercialize technology will find that the possibility of infringing others' patents will have an effect on their efforts. Accusations of infringement may come from a variety of sources—if not from competitors also seeking to commercialize the same or similar technology, then from those who have acquired the patents simply for their potential in generating licensing revenue.

There are good reasons to believe that commercialization of nanotechnology will be particularly susceptible to this kind of interference. The publicity surrounding nanotechnology is reflected in the formation of a relatively large number of start-up companies, most of which will be filing patent applications and most of which will ultimately fail as businesses. For the most part, the patent applications that they file, though, will not simply disappear with the failure of the companies. Instead, they will be sold as part of an effort to generate revenues or their ownership will simply pass to whatever organization acquires the start-up.

The result of this is that many patents that cover fundamental aspects of nanotechnology will be owned by those who acquire them simply for their licensing value. To realize a return on their investment in these patents, their owners will engage in a strategy of identifying possible infringers and asserting the rights granted by the patents to demand compensation. A company that is on the receiving end of an allegation of infringement and a demand for payment is likely to portray itself as having been victimized by a patent troll.

## ii.  Willful Infringement

While it is impossible to eliminate the risk of an allegation of infringement interfering with a business, there are a number of steps that can be taken to limit this risk. Perhaps most important of all is to be certain that there is never any willful infringement of a patent. Willful infringement occurs when someone is aware that activity infringes a patent and nevertheless continues to engage in that infringing activity. When the infringement is proved, it is punished by making the infringer liable for as much as three times the actual damages to the patent holder.[27] In cases where the infringer was not acting willfully, the liability is usually limited to the actual damages to the patent holder.

A determination of whether someone is acting willfully may depend on a huge number of different factors. The easiest way to prove willfulness is, of course, to find a document written by the infringer proclaiming that it knows its activity is infringing and yet plans to continue it anyways. In reality, such documents are never found so that proof of willfulness is built up by establishing various inferences. Fundamental to these is proving that the infringer was aware of the patent, and particularly that it was aware of the relevant claims. The most common way of establishing this is for the patent owner simply to send a copy of the patent to the potential infringer, keeping a copy of the post office's proof of delivery, and see whether the potential infringer persists in its activities.

How should a nanotechnology company respond when it is on the receiving end of a letter like the following?

Dear Mr. Smith:

Enclosed is a copy of our patent, U.S. Pat. No. X,XXX,XXX, which claims methods of using nanoscale ceramics in producing sunscreens. It has come to our attention that you may be using the methods claimed, particularly those of Claims 1, 7, 8, and 14. Please contact me at your earliest convenience to discuss the possibility of acquiring a license to the patent.

Sincerely,

Pat T. Roll

Such a letter should not be ignored out of hand since it likely represents the first step in a complex strategy to compel the business to hand over money—possibly in the form of license fees but more likely in the form of a settlement. In some cases, the level of analysis that needs to be performed might be very modest. For instance, suppose that the relevance of the claims to the company's business seem ridiculous—say the company uses nanoscale ceramics in the production of coatings for display monitors; its coatings are entirely unsuitable for skin applications and they have no intention of ever manufacturing a skin product. In such a case, verifying that the claims are indeed limited to sunscreen applications may be sufficient to be able to treat the letter like any other unwelcome solicitation: to ignore it. In such cases, the solicitation is much like a phishing strategy in which a scattershot approach is used to contact every company that mentions "nanoscale ceramics" on its Web site in the hopes a few will respond.

Other circumstances in which it is plain that some action might be needed are those where the company does, in fact, produce sunscreens using nanoscale ceramics. A superficial reading of the claims might seem to confirm the worst fears of the company's president—that notwithstanding its engineers' beliefs, someone does seem to have scooped them in coming up with the basic idea on which the company is founded. In those cases, it is usually necessary to involve a patent attorney since there are still a number of ways in which to respond.

Still other circumstances—probably the majority—fall somewhere in between, where it seems as though the company is doing something different than the patent claims but at the same time seems uncomfortably close. This might occur when the company does not manufacture sunscreens but does use the manufacturing method described in the patent to fabricate a variety of different kinds of cosmetics. Or perhaps the letter was being a bit disingenuous in characterizing the claims. It could be that the claims actually refer only to "porous nanoscale ceramics" when the company is using materials that have relatively low porosity. At what point does a "nanoscale ceramic" become a "porous nanoscale ceramic"? In these cases, it is prudent to involve a patent attorney to better understand what the precise scope of individual claims might be.

### iii.  Evaluating Claim Scope

While there are a number of subtleties that might be involved in the detailed application of patent law, it is possible to highlight a few basic considerations that a patent attorney will make. The first objective is to understand the precise scope of an individual claim. It is very often the case that when one first reads a claim, it seems to be written in very broad and general terms so that it has a large scope. But this is not the end of construing the meaning of the claim, and further readings tend to highlight the importance of specific terms.

In many cases, a patentee will have made statements that cause some of those terms to be interpreted more narrowly. Some of those statements might be in the written description portion of the patent, but others will have been made in the exchange with the patent examiner in trying to convince the examiner to allow the claims. All of the communications with the patent office are publicly available and may be used in the analysis of patents. Suppose, for instance, that while the claims refer to "nanoscale ceramics," every single example in the written description is of zinc oxide. And suppose further that to secure allowance of the application, the applicant repeatedly told the patent examiner of how there are specific difficulties in using zinc oxide that are not shared by other ceramics and that the method was specifically designed to address those difficulties. In such a circumstance, it might be appropriate to construe "nanoscale ceramic" as meaning "zinc oxide," permitting a company that uses titanium oxide to avoid infringement.

While the apparent scope of claims can sometimes be narrowed in this way, there are also mechanisms that can cause claims to be construed more broadly. Usually, this occurs with the *"Doctrine of Equivalents,"* which permits a patent owner to assert infringement against someone who has made only insubstantial modifications to an invention. The idea is that an infringer shouldn't be able to avoid liability by making tiny tweaks to an invention— substituting a screw for a nail or subtly changing other environmental conditions like pressure to allow the process to proceed at a different temperature. This ability to broaden the reach of claims is severely limited, but it is important to consider, for example, whether a facial night cream that uses nanoscale ceramics is "equivalent" to a sunscreen that does so.

A proper determination of the claim scope in this way permits a better determination whether there is actually infringement. In some cases where there does appear to be infringement, it is because the claims are so broad that they are not valid. This is often the case when the president shows the patent to the company's engineers who scoff that the claims cover what some research group was publishing eight years ago and that what the company is now doing is advanced beyond that in various respects. This advancement is not relevant to the infringement question, but the activity that was previously being performed raises questions about the validity of the claims.

It may be possible to find publications or other prior art that can be used to establish that the claims are not valid. While the knowledge of the company's own scientists and engineers is frequently a good source of leads for finding invalidating prior art, it is also often the case that professional searchers can find useful information. There are a number of experts available who will perform such services for reasonable fees. In cases where it is more difficult to find invalidating prior art, the importance of invalidating the claims may justify employing foreign searchers who will search patents and publications only available in other languages. Japanese patent publications and Russian technical publications are often good sources of prior art. In showing that certain claims are invalid over prior art, exactly the same considerations discussed in the section "Prior Art" above apply.

### iv. Opinions of Counsel

To protect the business from a finding of willful infringement, the patent attorney will frequently prepare a written analysis that sets forth the precise basis for her opinion. When the analysis is sufficiently well reasoned that the company is behaving reasonably in relying on it, it will generally be protected from a finding of willful infringement. The attorney's opinion does not guarantee that a judge or jury will not reach a different conclusion, so it is still possible for some claim of the patent to be infringed. But in such cases, the liability is usually limited to actual damage to the patent holder rather than treble damages.

Because of the requirement that the opinion be well reasoned, it is usually presented in the form of a detailed document that sets forth the precise legal and factual basis underlying it. While the length and intricacy of the analysis can sometimes be daunting, the requirement that the company's reliance be reasonable generally requires some effort to understand the nature of the arguments being made. One convenient way to meet this requirement, even if every nuance of the document is not fully digested, is to ask the attorney to make a presentation to key personnel in the company. The objective of such a presentation is to ensure that they understand the basic content of the opinion—should the need arise, they will be able to testify before a jury that they went through this exercise to show that their conclusion that the company does not infringe a valid claim was a reasonable one.

In many ways, obtaining an opinion of counsel is a relatively passive response to an allegation of infringement. The factual and legal position underlying such an opinion could instead be used in a much more aggressive fashion. In some instances, the ultimate effect of a more aggressive approach will be to remove completely any cloud that the infringement allegation presents.

### v. Designing around Patents

In some instances, for example, it may be possible to design around the patent claims. In going through the exercise of determining the precise scope of the claims, it may become clear that relatively simple changes in a product may completely eliminate any argument that there is infringement. The specific types of changes that might be made are as variable as the types of patent claims that might exist. They could be in the nature of simple structural changes to a NEMS device, a change in the chemical composition of some nanotechnology substance, a change in the way or order in which certain manufacturing process steps are performed, and so on.

Whether to implement the design change ultimately hinges on the relative cost of the change and the degree to which the risk of being found liable for infringement will be affected by the change. If the change will convert a clearly infringing product into one that is certainly not infringing, even a relatively high cost for the change may be worthwhile. But if the change has a more modest effect—say, changing the risk of being found liable for infringement from 60 percent to 40 percent—it might only be implemented if its cost was similarly modest. In many instances, it is still appropriate to have an opinion of counsel prepared that explains why the modified product does not infringe, particularly if the design change does not avoid all potential arguments of the patent holder. Such an opinion may still insulate the company from a finding of willful infringement should one of those arguments ultimately succeed.

## vi.  Reexamination of Patents

Even more aggressive responses include using procedures available in the patent office or in the courts to obtain decisions that prevent the patent holder from asserting the patent. The U.S. Patent Office provides a *reexamination* procedure in which third parties may request that the patent office reconsider the reasons that caused it to issue the patent in the first place. The person requesting the reexamination must provide an explanation why certain prior art raises a question whether the patent claims are actually valid. Depending on the nature of the request, the requester may either continue to participate or be prohibited from providing further input. The advantage of making the request and letting the patent office proceed on its own is that the requester may remain anonymous. If the requester wants to continue to participate, she will need to disclose her identity to the patent holder. While this procedure has the apparent advantage of being able to obtain a more balanced consideration by ensuring that it is not only the patent holder providing arguments to the patent office, in practice both procedures generally favor the patent holder.

## vii.  Declaratory Judgments

Another limitation of the reexamination procedure is that it is only available to attack the validity of the patent. The *"declaratory judgment"* procedure available in the court system provides mechanisms to have a court declare that the claims are invalid or that a particular product does not infringe. It is therefore considerably more versatile than the reexamination procedure. It is also more evenly balanced in the way it permits evidence and arguments to be presented by both sides. Perhaps not surprisingly, though, it is a considerably more expensive procedure. Sometimes this cost may be defrayed by cooperating with competitors who have also been accused of infringement—or are likely to be if the defining pattern of a patent troll is maintained. In the end, this type of coordinated response to a patent troll may prove to be the most effective.

### viii.   Patent Thickets

These various options for addressing potential infringement issues may be used in a variety of different circumstances. There is no need to wait until a patent troll has arrived at the doorstep. Indeed, as the lessons of many childhood folk tales attest, a decision to wait for a troll to make the first move may well be at one's peril.

In the context of nanotechnology, there are other reasons that may be even more compelling to prompt defensive responses to possibilities of infringement. Perhaps the most critical among these is the apparent development of *"patent thickets"* around nanotechnology. A patent thicket arises when the patent rights granted to different entities overlap in a significant way. The term appears to have originated in the mid-1970s in connection with litigation involving the domination Xerox was exerting over the photocopying industry.[28] More recently, the term has acquired more prominence and is described with evocative metaphors, like the definition provided by economist Carl Shapiro: "a dense web of overlapping intellectual property rights that a company must hack its way through in order to actually commercialize new technology."[29] Indeed, much of the criticism alluded to earlier that patent laws act to stifle innovation rather than promote it may be traced to the development of patent thickets in certain fields.

The clearest indication of the development of a patent thicket in nanotechnology is the recent sharp growth in the number of patent applications being submitted in this area (see Figure I.8). This growth is coupled with a general novelty of the field as a whole, making it more difficult for the patent office to find and evaluate prior art than is the case with more mature fields. The result is that many nanotechnology companies will find intricate webs of patent claims surrounding their particular niche.

The conventional strategy of entering into cross-licensing arrangements becomes much less attractive as the number of parties to be involved in such arrangements increases. Quite simply, the transaction costs in developing such arrangements become prohibitive. But a variation on this idea has often been found effective in addressing patent thickets, and is likely to be used in various nanotechnology areas. A *"patent pool"* is a structure that arises when different companies assign or license their patent rights to a coordinating entity. This coordinating entity then takes on responsibility for exploiting the collective rights provided by the various members of the pool. Companies are motivated to join the pool because they all face the same fundamental barrier presented by the patent thicket, with each of them contributing its own piece of that thicket.

There are a variety of different structures that may be used by the coordinating entity to control licensing of the collective rights. One common structure is centered on the development of an industry standard, with the patents included in the pool including claims that are determined to be essential to practicing this standard. It may well be the case that the development of standards in the various emerging nanotechnology industries may be spurred by the need to address the development of a patent thicket.

From the perspective of a nanotechnology company evaluating its options, joining a patent pool may prove to be relatively attractive. While it would then have to give up much control over the patent rights it acquired, membership in the pool would, at the same time, relieve it of a variety of concerns presented by the patent rights that others hold. By acquiring at least some form of right to practice the inventions embraced by those other patents, it may avoid the expenses associated with evaluating their validity, with attempting to develop alternative designs that avoid infringement, with launching and participating in reexamination proceedings in the patent office or in declaratory judgment proceedings in court, or with any of a variety of other potentially costly ways of addressing the patents on an individual basis. In short, the real value of its investment in obtaining patents may be in having earned membership in the patent pool.

## ix. Assessment

The weapons that patent ownership provides are potent. They provide a significant ability to interfere with the commercial activities of competitors. At one level, they can be used to impose a financial drain on a competitor by compelling the competitor to address a host of issues—evaluating validity of the patents, evaluating claim scope to determine infringement, developing alternative designs, and so forth. At a higher level, they can be used to obtain injunctions that will prevent the competitor from even marketing competing products. There is no doubt that to be successful, nanotechnology compa nies will have strategies for defending themselves against such weapons— in many cases developing their own patent arsenal as a deterrent under a theory of mutual assured destruction.

But the importance of patents to these companies must always be kept within the appropriate context. Again, the very earliest days of the patent system in the United States provide a cautionary tale—reflecting the eternal character of issues that confront those attempting to develop technology, whether it be nanotechnology or steam technology. Much of the very early history of the patent system in the United States was centered on different inventors seeking to obtain patents for steamboats and uses of steam. These different inventors sought to obtain an advantage over the others by establishing their rights to patents in aspects of the technology. The efforts they went to were considerable, and descriptions of the personal toll their efforts took make for intriguing and colorful stories.[30] In the end, many of them were awarded patents. But it wasn't until well after the earliest of these patents had expired that Robert Fulton was finally able to make the financial arrangements that would bring the conception of a steamboat to practical reality. Will nanotechnology follow a similar pattern? Only time will tell.

### Discussion

1. On May 15, 2006, the Supreme Court of the United States ruled in *eBay v. MercExchange* that finding that a patent has been infringed should not necessarily result in a permanent injunction that stops the infringing activity. This finding has been described as a victory against patent trolls since it appears to limit their ability to interfere with a business's operations. But the ruling applies equally to suits brought by those who own patents but produce no products as is does to suits brought by active competitors in the marketplace. Evaluate the impact of this ruling on how patents affect the ways active manufacturers compete in the marketplace. Is this effect positive or negative? How is the dynamic in licensing negotiations likely to be affected? Is this ruling consistent with the fundamental idea of patents providing a "right to exclude others" from practicing the invention? After considering these issues, you may wish to read the Supreme Court's opinion at *eBay Inc. v. MercExchange, L.L.C.*, 547 U.S. 206 (2005) to see how your views compare with those of the various Supreme Court justices.

2. I previously noted that the United States has a *"first-to-invent"* system that may provide superior patent rights over the first to have filed a patent application. The procedure by which these superior rights are acquired is an *"interference"* proceeding (with the name reflecting the fact that the rights claimed by different parties interfere with each other). Research the requirements for initiating an interference proceeding in the United States and what is involved during such a proceeding. To what extent might prompting an interference be a reasonable response to an infringement allegation? What advantages and disadvantages do you see such a response having over other types of responses? Your response should specifically address ways in which interferences are similar to or different from other patent office proceedings like reexaminations, and ways in which they are similar to or different from court proceedings like declaratory judgment actions.

## 5.  Nanotech Patents outside the United States

From the perspective of a patent holder, one of the most significant limitations of the patent system is its geographic restriction. A United States patent gives the patent holder certain rights to exclude activities by others in the United States. And similarly, a Chinese patent gives the patent holder comparable rights in China. A Russian patent gives the patent holder analogous rights in Russia. And so on. But what does not exist is the ability to hold a single patent that confers rights in all countries. Like many emerging technologies, nanotechnology is likely to find application in an international context. Companies developing nanotechnology would ideally like to have rights conferred in many countries, notably in various European and Asian countries, as well as in North America.

### i. The Paris Convention

The idea of an *"international patent"* is one that receives attention from time to time. But at least so far, there remain sufficient differences in the way that different countries approach patent issues that no such instrument has developed. Instead, there are a number of treaties between countries that attempt to provide at least some rights in a second country derived from the filing of a patent application in a first country. At its core, these right are grounded in the Paris Convention, which was first signed in Paris on March 20, 1883. It was initially signed by eleven countries: Belgium, Brazil, France, Guatemala, Italy, the Netherlands, Portugal, Salvador, Serbia, Spain, and Switzerland. Now almost every country in the world is a signatory to the convention.

What the Paris Convention essentially does is to give effect to the filing date of a patent application in any of the signatory countries, as long as another application is filed within a year. That is, an application filed in the United States (or in any other signatory country) may be used as a basis for an application in Germany, Australia, Japan, or any other signatory country within that one-year window. This at least provides a mechanism by which the potential for patent rights in most countries of the world can be preserved by a single filing.

### ii. The Patent Cooperation Treaty

The Paris Convention is augmented by the Patent Cooperation Treaty (PCT), which came into force in 1978. It has been signed by the majority of the Paris Convention countries, but there remain a few countries that are party to the Paris Convention and yet not party to the PCT. Before the existence of the PCT, the one-year anniversary of the filing of a patent application marked the date at which a decision was needed for filing counterpart applications in other countries. If the cost of filing an application in each country is more or less commensurate with the cost of filing in the United States, then the cost is multiplied by the number of countries of interest. This can easily mean that seeking patent protection for a nanotech invention in any sort of international fashion can cost hundreds of thousands of dollars.

There is no way ultimately to avoid this cost if international patent protection is to be had, but the PCT does provide an effective mechanism for at least deferring the bulk of the cost. Such deferral provides additional time during which a company may decide whether the nanotech invention is of sufficient commercial interest to justify the high cost of diversified international protection. In some respects, a purist view of the PCT would not focus on this deferral aspect but would instead emphasize the various procedural benefits that are available under the PCT to consolidate certain parts of the examination process. But the practical—and unquestionably most widespread—use of the PCT is as a mechanism for delaying the need to file counterpart applications in other countries.

The ability to defer the time of filing in individual countries is a consequence of the fact that the PCT provides a structure that allows for the consolidation of parts of the search and examination of patent applications. This takes time, with the resulting time structure being defined to have a total of eighteen months. The deadline for filing in individual countries (*"entering the national phase"* when such an application is based on a PCT application) is accordingly thirty months from the date the application was first filed in one of the PCT countries. Some countries even extend the deadline to be a little bit longer than the nominal PCT deadline, with European and several other countries providing a total time of thirty-one months. (Figure I.10 shows a timeline of the PCT application process.)

The processing of a PCT application may be handled by any of a number of offices located in different parts of the world, reflecting the international character of the PCT. Each of these offices is a patent office that has experience in the examination of applications, and each office may act in a different capacity at different parts of the overall PCT process. For example, the office may act as a "receiving office" that is responsible for accepting PCT applications and ensuring that certain formalities of the application are satisfied—that it includes a written description of the invention, a set of claims, suitable drawings, properly executed filing documents, and so on.

Applicants are not generally free to choose which receiving office to use for submission of an application. Rather, there are certain geographical limitations requiring that particular receiving offices be used for applicants based on their citizenship and/or residency. In addition to the various patent

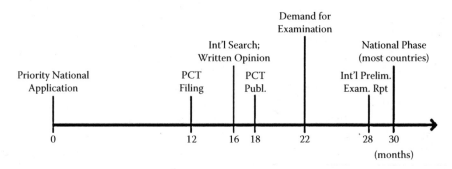

**FIGURE I.10**
Example PCT timeline.

offices that may be used as receiving offices, there is an International Bureau of the World Intellectual Property Organization located in Geneva that may accept applications from any applicant who is a national or resident of a PCT country. The International Bureau also provides a number of other support functions in administering the PCT and receives a copy of all the applications filed with any of the receiving offices, acting as the body that maintains the master files. For instance, the International Bureau is responsible for publishing the application at the appropriate time and acts as a general coordinating body for all PCT applications.

All applications filed under the PCT are subject to an international search by one of the patent offices acting in its capacity as an "international searching authority." Such a search is similar to the search that would be performed by that patent office when searching for prior art. An international search report is issued identifying particular references according to their perceived relevance to the patentability of the claims, and is transmitted to the applicants with a written opinion. An opportunity is provided at this stage of the process for the applicants to take any of a number of different actions, including withdrawing the application or amending the claims in light of the references uncovered by the search. As long as the application is not withdrawn, the application is published by the International Bureau with the international search report (but not with the written opinion). If the claims are amended, a second written opinion may sometimes also be issued.

In addition to receiving a written opinion based on search results, it is possible to have an examination of the application performed by an international preliminary examination authority. Such an examination is only performed upon the filing of a special demand for such examination and the payment of an additional fee. Such an examination can be of value because the results of the examination may be accepted by many of the patent offices around the world when the PCT application enters the national phase and is filed in individual countries. It should be noted, though, that not every country in the world accepts the international preliminary examination report, with those countries that have relatively sophisticated patent offices being generally less likely to accept the results of examination by a different examination authority.

Examination of the application during the PCT process affords another opportunity for amending the claims prior to entry into the national phase. Between the search and examination, there are thus multiple opportunities to consider prior art found by a patent office and to put the claims in better condition for allowance. In cases where an application is likely to be filed in many countries, this type of consolidation of at least much of the search and examination can be particularly valuable. The claims can be cast into a form that addresses all the issues raised by the search and examination, making it ripe for an easy allowance in those countries that defer to the examination of the examination authority.

Once these procedures have been completed, the application is ready for entry into the national phase. This is the point in the process at which the consolidation provided by the PCT is completed. The application is then filed within each country of interest according to the specific criteria for that country. In some respects, this last statement is not entirely accurate in that there are certain parts of the world where another level of consolidation may be used. Undoubtedly the most commercially important of these is Europe, in which the European Patent Office will examine an application on behalf of all member states. In this instance, it is not necessary to have a separate examination performed in the German, French, and Spanish patent offices when patents are desired in those countries. Instead, a common examination is carried out by the European Patent Office, requiring only an administrative validation procedure after the application has been allowed to bring patents into force in the identified countries.

## Discussion

1. While the United States requires that claims define inventions that are novel and nonobvious over the prior art, most other countries of the world require that they be novel and define an *"inventive step."* Research how these standards differ.

2. Under the provisions of the PCT, applications filed in the U.S. receiving office may be searched by either the U.S. Patent Office or by the European Patent Office acting as an international searching authority. What considerations do you think might apply in selecting one searching authority over another? Do you think the relevant considerations are different for nanotechnology applications than for applications in other technology areas?

3. Imagine you are the chief technology officer of a start-up nanotechnology company. There are six technical scientists and engineers who generate a total of about forty inventions a year. Assume every invention is patentable. For the purposes of this exercise, each U.S. provisional application (see Discussion topic 5 under the section "Riding the Patent Office Pony") has a cost of $2500, each U.S. nonprovisional application has a cost of $10,000, each PCT application has a cost of $5000, and each national phase application has a cost of $5000 (irrespective of country). An award of $1000 is given to the inventors for each disclosure used as a basis for a patent application. Your patent budget is $250,000. Your board of directors wants broad, effective patent coverage over as much of the world as possible. What patent filing strategy do you use? How many U.S. applications do you file? Do you file them as provisional or nonprovisional applications? How many PCT applications? How many national phase applications? Have a partner challenge your strategy from the perspective of an inventor whose invention is rejected, from the perspective of the board of directors, and from the perspective of your chief executive officer. Are you able to defend your strategy effectively?

## B. Copyrights

### 1. Introduction

When one looks at images of geodesic domes, like that pictured in Figure I.11, it is impossible not to be struck by their artistic beauty. Geodesic domes are formed from an array of mechanical struts that create a network of triangular elements. These triangular elements have tremendous functional value to the structure, distributing stress across the structure in a way that gives it great strength. Indeed, this property makes geodesic domes the only manmade structures that get proportionally stronger as they increase in size.

At the same time, this network of triangular elements folds over on itself to form an almost spherical structure that is remarkably pleasing to the eye. Geodesic domes have the unusual characteristic that they simultaneously look like naturally occurring and manmade structures; while the structure as a whole resembles honeycombs that appear in nature, the mechanical engineering of the structure is somehow always evident. In this way, geodesic domes are a prime realization of perhaps the most fundamental goal of architecture—to combine utility and artistry in a manmade construction.

When the $C_{60}$ molecule was discovered in the mid-1980s, it was inevitable that the same type of observation would be made. Here was a structure created by nature that somehow also looked as though it had been engineered. The christening of this structure as *"buckminsterfullerene"* was effective as an

**FIGURE I.11**
An image of a geodesic dome.

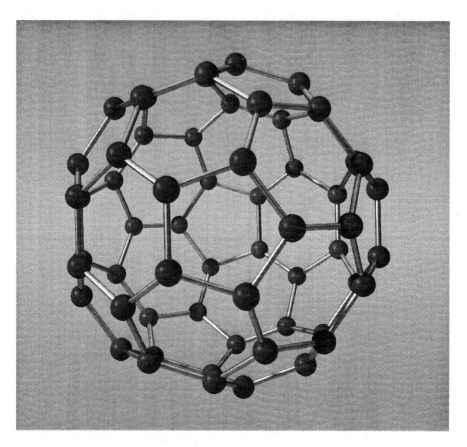

**FIGURE I.12**
An artistic representation of buckminsterfullerene $C_{60}$.

homage to the inherent duality of architecture in seeking structures that are both functional and artistic. Named after R. Buckminster Fuller, the architect most famous for developing the ideas of geodesic domes, the $C_{60}$ molecule is a functional structure that is also striking in its artistic beauty (Figure I.12).

The basic $C_{60}$ structure is still often referred to informally as a "buckyball" after the nickname by which Buckminster Fuller was known to his close associates, "Bucky." In many ways, the buckyball remains an archetype for nanotechnology itself, being the basis for the development of numerous technical ideas in nanotechnology. It also seems that its persistent archetypal status may well be a result of the way that it combines aesthetics with function—there are aspects of nanotechnology that are very much like large-scale architecture in embracing a duality of functionality and artistry.

The more generic term *"fullerene"* is, of course, a clipped form of *"buckminsterfullerene."* It refers to a wide variety of structures that may be formed from carbon atoms, including both closed structures like that shown in Figure I.13 and tubular structures like that shown in Figure I.14. These structures also

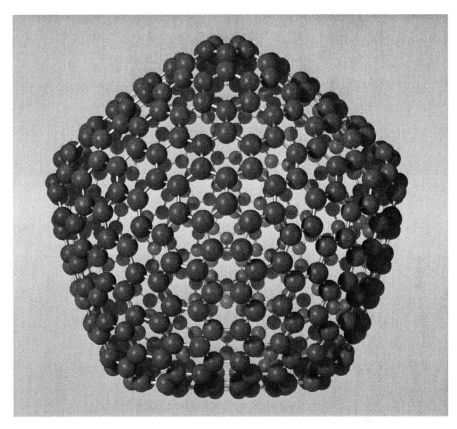

**FIGURE I.13**
An artistic representation of fullerene $C_{540}$, an example of a large closed-shape carbon molecule.

have a visual impact that makes their structure appear to have an aesthetic component. Yet, at the same time, these structures are among the simplest of nanotechnology structures. Considerably more complex structures have already been designed and constructed and there is every guarantee that future designs will have yet more complexity and more aesthetic appeal. In many cases, including such an aesthetic component will be part of an intentional design—many nanotechnology engineers will approach their creations in a manner similar to the style adopted by architects. When there are multiple ways in which desired functionality can be achieved, it will be preferable to seek the design that has aesthetic appeal.

Intellectual-property law provides different mechanisms for protecting the functional and artistic aspects of ideas. Patents are by far the most important mechanism available for protecting functional aspects of inventions. Indeed, as an architect, Buckminster Fuller obtained twenty-eight U.S. patents over the course of his career, many of them directed to geodesic domes and similar structures. Protection of the artistic aspect of ideas is generally provided

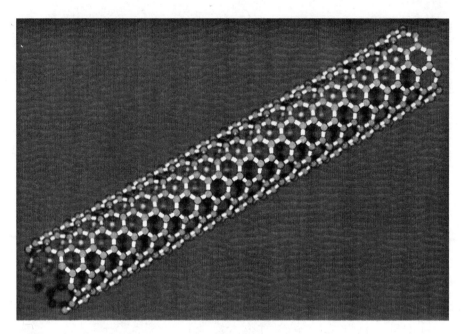

**FIGURE I.14**
Artistic representation of part of a carbon nanotube.

by copyright law. The copyright laws of many nations include specific provision for the protection of "architectural works."

To what extent might copyright law provide protection for the creation of nanotechnology structures? The previous discussion alludes to parallels that exist between architectural structures, which are often afforded copyright protection, and certain nanotechnology structures. But the comparison between large-scale architecture and nanotechnology is not a precise one. There is more that distinguishes a buckyball from a geodesic dome than simply scale.

Probably most important in considering the potential for copyright protection when applied to nanotechnology is that all of the structures shown above are artistic renderings of nanotechnology structures and do not show the precise structures. The ball-and-stick model of molecules is very useful in conceptualizing molecular structures, but that is not how actual molecules appear. At the scale of molecules, there are a variety of quantum-mechanical effects taking place that affect how molecules can be observed. The individual atoms are not hard spheres, but are a complex and ever-changing jumble of more fundamental particles. And the bonds are not rigid sticks between the atoms, but are collections of electrons that have convoluted orbits that permit them to be described as "shared" by the atoms.

The term *"nanotechnology"* is increasingly used to describe a range of scales, and the ability to discern structure in a conventional sense may well depend on the precise scale of individual structures. Structures on the scale of a hundred nanometers can readily be resolved in a traditional microscopic sense and should be entitled to copyright protection for certain artistic characteristics. The situation becomes less clear as the scale continues to decrease. At some point (and perhaps it is when the term *"picotechnology"* begins to be in vogue), quantum-mechanical effects make it impossible to image structures in any conventional sense. The ability of copyright protection to apply to those kinds of structures becomes significantly limited.

### Discussion

1. The analogy between fullerenes and geodesic domes is a natural one. Can you think of any other effective analogies between nanotech structures and macroscopic structures? What weaknesses exist in those analogies? Try to keep those analogies—and those weaknesses—in mind when considering the requirements for obtaining copyright protection discussed in the next section. How do you think copyright would apply differently to the macroscopic structures and the analogous nanotech structures?

## 2. Copyright Requirements

The general requirements for claiming copyright are relatively easily stated: fixation, originality, and creativity. Some form of these requirements exists in all copyright systems of the world.

### i. Fixation

*Fixation* represents the fundamental condition that the doctrine of copyright provides protection only for ideas that have been memorialized in some tangible fashion. This acts to distinguish between ideas by themselves and expressions of those ideas. By extending copyright protection only to the tangible expression of ideas, the law imposes a threshold level of development that must be met—ideas that remain in a creator's head or have only been communicated orally cannot enjoy the protections that copyright provides. It is only once they have been expressed in a way that they can be perceived by others in something other than a purely transitory nature that they cross the threshold.

The law is not at all restrictive on the form that the tangible expression must take. It is this flexibility that permits new kinds of structures—like those that might be made at a nanometer scale—to be entitled to copyright protection. In addition, provided that the other copyright requirements are met, it is the moment at which the expression becomes "fixed" that the copyright becomes established. This is a significant difference from patent rights, which require the satisfaction of considerably more formalities in order to be established.

There is, in particular, no requirement that any government body be notified of the work; once the expression is fixed, its creator has the right to identify it as subject to copyright protection. The law does often provide a number of inducements to encourage the registration of copyrighted works, the most notable in the United States being the requirement that there be a registration with the copyright office before an infringement suit can be filed. But this registration may be made at any time during the life of the copyright; the registration of copyrights does not implicate the same kinds of notice functions that the issuance of patents do.

### ii.   Originality

The *"originality"* requirement expresses the prerequisite that a copyrighted work must be the original production of its creator. The threshold for originality is very low and means only that it was not copied from another. Wholly new creations thus easily meet this requirement, even if they are similar—even very similar—to other preexisting creations. Where the originality requirement becomes most important is where a creation combines some copied material with some material newly developed by its creator.

Such a *"derivative work"* is entitled to copyright protection as its own creation, even as some of the underlying components of the derivative work are themselves subject to copyright. With such derivative works, the assignment of rights may sometimes have similarities with the apportionment of patent rights where different parties have patents with claims of different scope. For example, consider a derivative work created by Party B using materials created by Party A and protected by copyright. To create the derivative work, Party B must have permission of Party A to reproduce the underlying portions owned by Party A; but Party B owns the new material added in creating the derivative work and the derivative work as a whole (Figure I.15). Someone wishing to reproduce the derivative work would certainly need the permission of Party B, and if no provision had been secured by Party B to grant permission to reproduce all the underlying pieces, permission of Party A would also be needed.

### iii.   Creativity

The *"creativity"* requirement also has a very low threshold that must be met. To be entitled to copyright protection, there must be some minimal element of creativity in the work. This requirement has traditionally been used to exclude rote compilations—such as an uncategorized telephone book or a list of parts in a structure to be assembled—from enjoying copyright protection. In the context of nanotechnology, this requirement might form a basis for requiring some intent on the part of the creator to produce the nanotechnology structure. Structures that occur spontaneously at nanometer scales might be excluded from copyright protection as lacking creativity.

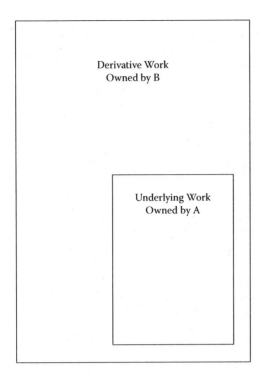

**FIGURE I.15**
Illustration of a derivative work.

### iv. The Scope of Copyright

Once a person has established a copyright, there are a number of rights that he can exercise. In a manner similar to patents, these rights are "exclusive" in that they permit the copyright holder to exclude others from engaging in certain activities. Probably the most fundamental right granted to a *copy*right holder is the right to exclude others from making reproductions of the work. In addition, though, the copyright holder may exclude others from preparing derivative works and from distributing copies of the work. Depending on the type of work, there are also provisions for excluding others from public performances or displays of the work.[31]

Also similar to patent rights is the fact that the rights granted to a copyright holder are of limited duration. But in the case of copyrights, the term is considerably longer: in the United States, the copyright in a work expires only seventy years after the death of the creator. Works created by young individuals can easily persist for well over a hundred years.

Someone's copyright is infringed when one of these restricted activities is performed without the permission of the copyright owner. The potential penalties are varied, ranging from the payment of modest financial damages to seizure of the infringing work and potential imprisonment under criminal statutes—as anyone who has ever rented or purchased a movie has duly been warned.

The breadth of possible responses to copyright infringement is entirely appropriate. It is a reflection of the fact that lines are not always easy to draw when assessing creative works. If something is an exact copy of a creative work, it is almost certainly a knockoff and its theft should be punished appropriately. But in most cases, copies are not exact. Instead, there are a variety of ways in which some new work is similar to some preexisting work and a variety of ways in which they differ. When has the later work merely been "inspired" by the earlier work but includes its own creative content, and when has the creative content been taken wholesale from the earlier work with the addition of superficial cosmetic changes to make it "different"? How are the meaningful creative and noncreative portions of works identified, and who decides where to draw these lines?

These are not easy questions. The law provides a form of compromise, though, by explicitly authorizing the *"fair use"* of copyrighted material. There are four factors that are to be assessed in deciding whether a particular use is "fair" (Figure I.16).[32] This structure does not remove the need to exercise judgment when considering copyright issues, but it does provide a structural framework within which that judgment should operate.

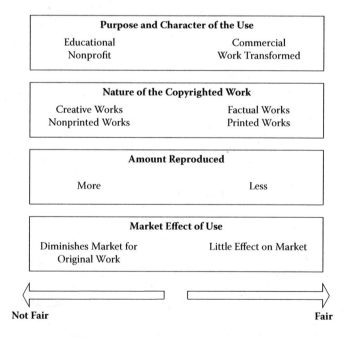

**FIGURE I.16**
Factors involved in determining whether a use of copyrighted material is fair.

### v. "Fair Use"

The first factor of fair use considers "the purpose and character of the use," with greater excuse being given to uses that are for educational and/or non-profit uses. This factor reflects one of the goals of copyright law as attempting to enrich the public by stimulating creativity. If someone else's creation is used for commercial gain, that use is less likely to be considered "fair" than nonprofit or educational uses. Also, if the original work is somehow "transformed" when it is used, rather than merely reproduced, that fact will also argue in favor of finding that the character of the use makes it "fair."

The "nature of the copyrighted work" is also relevant and acts as the second factor. The distinctions among different types of creations embraced by this factor again tend to reflect the objective of copyright law as encouraging creative production. The use of factual works is thus more likely to be viewed as "fair" than are the use of creative works. There is also a tendency to view the use of printed works as more "fair" than the use of nonprinted works.

Someone whose entire creation is reproduced is naturally more inclined to feel cheated than someone who has only a portion of her creation reproduced. This natural stance is reflected in the third factor, which considers the "amount" of the work that is reproduced. This consideration is made in relative terms by considering what portion of the entire copyrighted work is reproduced, rather than focusing on any attempt to define an absolute standard. This factor may sometimes pose practical difficulties because it is sometimes the case that the heart of a work is a relatively small portion of the whole.

The final factor considers the "effect of the use upon the potential market for or value of the copyrighted work." In some ways this factor is the most important, and in other ways it is the most difficult to apply. It is entirely sensible to define the use of someone's copyrighted work as less "fair" when it diminishes the value of the original creation or deprives that creation of a portion of its market. But as a practical matter, it can be very difficult to quantify these types of effects.

### vi. Where Nanotechnology Fits

What is evident from both the criteria for obtaining a copyright and from the fair-use defenses to copyright infringement is the fact that copyright protection is strongest for creative works. This is perhaps the most important distinction between (utility) patents and copyrights—what patents protect is largely the functional aspects of an invention. In this sense, these two intellectual-property doctrines are essentially complementary. The functionality of some invention is beyond the purview of copyright protection in the same way that the artistic or creative aspects are outside the scope of patent protection.

The question thus remains: which is the more important form of protection for nanotechnology? Like most responses that attorneys provide, the answer is "It depends." It depends on whether the functional or artistic aspects of some nanotech creation are more important. By far, the current emphasis on nanotech research is on functional structures that should be protected by patents. This is true of the majority of products that have already been commercialized, from surface coatings used in ski wax to tennis racquets that incorporate nanotubes.

But it would be a mistake to dismiss the potential that copyright protection provides. Although they are so small that they cannot be seen by the naked eye, there may yet be artistic aspects to nanotech structures that are amenable to copyright protection even when patents cannot apply. In the previous section, some examples were given of structures that have hybrid aspects of functionality and creativity. In the next section, the idea that nanotechnology may be a form of art is explored further.

### Discussion

1. The terms for patents and copyrights are vastly different. Why do you think this might be so? Research the historical trend for copyright and patent terms in the United States. Are the trends similar or different? What factors might account for the similarities or differences in these trends?

2. Designs. The "Patents" section focused on "utility" patents, which protect inventions that are considered to be useful. In the United States, there is another class of patents, called "design" patents, which protect ornamental aspects of inventions. These are examined by patent examiners in the U.S. Patent and Trademark Office in the same way as utility patents and are largely subject to the same kinds of prior-art restrictions. Many other countries have an intellectual-property doctrine that is completely separate from their patent systems and covers industrial designs. The protection that they provide roughly corresponds to the protection obtained with design patents in the United States.

   Research the criteria for obtaining design protection in the United States and elsewhere. Compare the protection that those doctrines provide as an alternative to copyright protection. Factors that you may consider include the relative term lengths, the required level of formality in obtaining the protection, and the ease of proving infringement. Based on your understanding of the criteria, can you think of any examples of nanotechnology structures that are entitled to design protection but not copyright protection? What about examples that are entitled to copyright protection but not design protection?

### 3. Nanotech Creations as Artistic Works

It may seem a bit unusual to think of nanotech creations as artistic works. After all, an entire gallery of nanotech creations could easily fit on the head of a pin and be crushed by a housefly. This makes it difficult to imagine any effective way of presenting nanotechnology as art in a conventional sense. There are, however, a number of artists who create conventionally sized representations of nanotech structures. These artists use a variety of techniques in approaching this task. Some produce paintings or sculptures based purely on their mental conceptions of nanotechnology. In some cases, these paintings are abstract, in the sense that they do not attempt in any way to reproduce how the nanotech structures actually look, but instead attempt to convey the mental and emotional impressions that the artist has of nanotechnology.

### i. Conventional Art

Other artists produce creations that do retain some impression or semblance of the appearance of nanotech structures, but subject to the creative input of the artist. One artist who takes such an approach is Cris Orfescu. Born in Romania in 1956 and holding a graduate degree in materials science, he currently lives and works in Los Angeles and has received recognition for producing what he describes as *"nanoart."* He often uses a process in which he takes electron-microscope scans of nanotech structures, using a computer-painting process, and then prints the result with archival inks on canvas. The result is a work that retains a visual impression of the nanotech structure complemented by the view of the artist (Figure I.17).

There is no question at all that these conventionally sized interpretations of nanotechnology, whether they are pure creations of the artist or based on

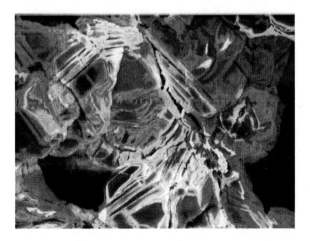

**FIGURE I.17**
A reproduction of "Power" by Cris Orfescu. This was created from an electron-microscope image of graphite nanoparticles. It is easy to detect the nanoparticle structure in this creation, even while getting a sense of the artist's conception of the nanostructure. (Image reproduced with permission of the artist.)

observations of actual nanotech structures, are entitled to copyright protection. But what about the nanotech structures themselves? The fact that Cris Orfescu uses electron-microscope scans reflects the fact that these structures are so small that they cannot be seen with visible light—the wavelength of visible light is hundreds of nanometers, and therefore too large to discriminate the much smaller nanotech structures.

### ii.  Nanosized Sculpture

In 2001, a group of researchers at Osaka University sculpted a bull out of resin using a two-photon micropolymerization technique—essentially using a pair of laser beams to act as tiny chisels to carve the image. Much like Michelangelo setting free the angel from the marble, the Osaka scientists released the tiny bull from the resin. This sculpture stands about 7 μm tall and has a length of about 8 μm. While some might quibble that it is thus a bit too large to qualify as a nanotech sculpture, it does serve as an example of a purely artistic structure that people are inclined to make at a very small scale. But even at this scale, it is still too small to be imaged with visible light, and the image of the bull provided in Figure I.18 was also obtained with an electron-microscope scan.

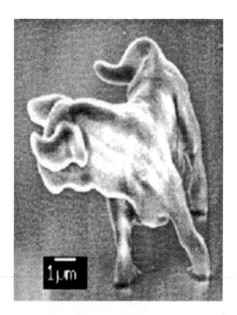

This is relevant in considering the applicability of copyright law to nanotechnology because the Osaka Bull is an example of a structure that has no utilitarian aspect at all. In addition to meeting the requirements of fixation, originality, and creativity, a work qualifies for copyright protection only when it falls within one of several defined categories.[33] Of the various defined categories, *"sculptural works"* are those that have the clearest relevance to nanotechnology. An object qualifies as a "sculptural work" only if it has a design of artistic craftsmanship that can be

**FIGURE I.18**
The Osaka bull. (S. Kawata, H.-K. Sum, T. Tanaka, K. Takada, Fine Features for Functional Microdevices, *Nature*, 412, 697–698 [2001]. Image courtesy of the Nature Publication Group.)

identified separately from the utilitarian aspects. Because the Osaka Bull is not at all utilitarian, it is easy to state without reservation that it qualifies for copyright protection. Anyone who produces a copy of the Osaka Bull without permission is liable for copyright infringement.

**FIGURE I.19**
An image of "NANO USA" inscribed by IBM researchers on a sheet of copper using 112 carbon monoxide molecules. (Image reproduced with permission of ASTM International.)

It is notable that the production of the Osaka Bull was not an isolated event. In the spirit of competition that frequently exists among scientific groups, researchers at IBM responded by using the technique of molecular-beam epitaxy to inscribe "NANO USA" and "IBM" with carbon monoxide molecules on copper sheeting (Figure I.19). Molecular-beam epitaxy is a process often used in nanotechnology to control the placement of individual atoms or molecules. After causing a molecule to stick to a surface, it can be moved by using *"photonic tweezers"*—again a pair of lasers, but in this instance used to apply light pressure to move the molecules.

Indeed, there have been a number of reports of nanotechnology companies now attempting to exploit a symbiosis between scientists and artists. The idea of scientist–artist collaborations is not a new one; there have been examples of such collaborations throughout history. The value of such collaborations is that artists and scientists often have different ways of looking at the physical world. By providing an environment in which they may exchange their views synergistically with each other, they are able to make much greater advances than either could alone. Indeed, some of the most significant achievements in history have been made by people who were accomplished both as scientists and as artists, Leonardo da Vinci probably being the most renowned.

Recognizing this potential for synergy, a number of nanotechnology companies have begun employing artists to work alongside scientists. Under such arrangements, it seems inevitable that at least some nanotechnology structures that they develop will include artistic features that may be distinguished from the purely utilitarian aspects of the structures. It is these aspects of the structures that will be entitled to copyright protection as structural works.

What about other categories? There are, of course a significant number of other categories of works that are entitled to copyright protection in addition to sculptural works. It is interesting to speculate whether any of these other categories might provide a basis for asserting copyright for nanotech creations.

### iii.  Architectural Works

Perhaps the most obvious category to consider, particularly in light of the earlier comparison between geodesic domes and fullerenes, is *"architectural works."*[34] It turns out that this category is easy to dismiss out of hand. The Copyright Act defines an "architectural work" as "the design of a building as embodied in any tangible medium of expression, including a building, architectural plans, or drawings."[35] While geodesic domes are frequently used as buildings, and therefore having designs that may be entitled to protection under copyright law, there is no question that nanotech creations are never used as buildings. To the extent that it is the structure of a nanotech creation that is to be protected, it is much more effective to view them as sculptural works than as architectural works.

### iv.  Literary Works

A category that may have more success in certain contexts is *"literary works."*[36] At first blush, this is rather surprising. When one thinks of "literary works," what comes to mind most immediately are great novels and other works of fiction—*The Grapes of Wrath, Gone With the Wind, The Sun Also Rises*, and so forth. But the Copyright Act defines "literary works" considerably more broadly as "works ... expressed in words, numbers, or other verbal or numerical symbols or indicia, regardless of the nature of the material objects ... in which they are embodied."[37]

What is of particular interest is that the breadth of this definition has provided a basis for copyright protection of computer programs and databases. As nanocomputers and nanoprogramming continue to be developed, copyright protection will extend to the programs and data that are used to control them. Futurists commonly classify nanocomputers into four categories: electronic, mechanical, chemical, and quantum nanocomputers. Of these, electronic and mechanical nanocomputers are considered to be miniaturized versions of more conventional types of computers. It is generally well known that the physical scale at which transistors can be produced has been decreasing by about a factor of two every eighteen months for decades now ("Moore's law"). This has permitted the development of microcomputers, which are now widely available. Further reductions in scale will bring these kinds of devices from the micro realm into the "nano" realm.

But what is perhaps more intriguing is the way in which chemical and quantum nanocomputers operate. A chemical nanocomputer stores and processes information in terms of chemical structures. In many respects, manmade chemical nanocomputers will merely attempt to duplicate what biological systems have already achieved. The individual instructions for constructing every living organism are contained in that organism's DNA genetic code. As Bill Gates is said to have observed, "DNA is like a computer program but far, far more advanced than any software we've ever created." The programming language of DNA is based on strings of four chemical building blocks called *"nucleotides"*: adenine, cytosine, guanine, and thymine. The appropriate sequence of these building blocks can define how to build everything from plankton to pterodactyls.

The fact that biological systems have been able to devise (remarkably compact) chemical instruction strings that define how to build such a diverse array of organisms offers hope that human beings will be able to write their own computer programs in chemical language. The sequences that they develop will be subject to copyright protection. Indeed, it is already the case that public attention has been drawn to the idea of individuals asserting a copyright over their own DNA. This attention has tended to come in the form of preying on common misconceptions over cloning to raise fears that celebrities and others may be duplicated without their permission. The scenarios that are raised are the stuff of science-fiction nightmares.

But what these efforts do highlight is the basic fact that chemical sequences are entitled to copyright protection as literary works. As research continues into the development of chemical nanocomputers, such sequences will be developed and protected. Some of this research will be in biological areas focusing on developing sequences that may be used to treat diseases in humans and other animals and used to produce desirable traits in edible crops and livestock, among a host of other uses. But other research will not be limited to the four nucleotides of major biological interest. There are countless other chemical structures that can be used in the generation of nanoprograms that may find other uses. It is quickly overwhelming to consider the possibilities that these present, and copyright law will provide one of the important forms of protection available to nanotech software developers.

The same basic analysis is true of quantum nanocomputer programs. The basic difference between quantum computers and more conventional electronic computers is the way in which information is represented. Electronic computers are based fundamentally on the use of binary information, representing all data and programming instructions as a sequence of 0's and 1's. Quantum computers may use such physical states as the spin of electrons to provide a similar binary representation. But quantum computers may extend this basic idea by using other quantum states that can efficiently represent more complex bit arrangements—a single quantum state could not only represent a single bit, but could represent two, four, eight, or sixteen bits. Programs expressed as a sequence of quantum states in defining programming instructions in this way would also be subject to copyright protection.

### v.  Musical Works

It is also interesting to speculate on the extent to which nanotechnology may be expressed as *"musical works"* or *"sound recordings,"* which are other categories that are entitled to copyright protection.[38] Interestingly, the term *"musical work"* is not defined in the U.S. Copyright Act, although a definition is provided for a *"sound recording"* as "result[ing] from the fixation of a series of musical, spoken, or other sounds."[39] A definition of *"musical work"* itself was not included in the statute because of Congress's belief that the term had a "fairly settled" meaning. What is clear is that, like literary works, there is no specific form in which the musical work must be fixed so that it is possible for it to be expressed in a variety of forms.[40]

To what extent can music be created by nanotech structures? This is not an easy question to answer. It is true that miniaturized versions of instruments have been created, like the nanoguitars that have been made at Cornell University (Figure I.20). Each of these guitars was fabricated by carving it out of silicon and the strings have a thickness of only about 50 nm. This focus on stringed instruments is a practical one because it is easier to generate actual acoustic vibrations from the strings than it would be with wind-type instruments. The Cornell researchers have actually played the 2003 version of the guitar by focusing laser light on the strings to set them in motion. The resulting sound has a frequency of about 40 MHz—roughly 2000 times the highest frequency detectable by the human ear and about 17 octaves higher than the frequency produced by a conventionally sized guitar.

Sequences of sound—music—produced by such nanoscale instruments should be as protectable under copyright law as any other type of music, notwithstanding the fact that their compositions would be outside the normal hearing range. It is also possible that music produced by nanoscale devices may have some practical applications. The frequency at which any string vibrates may be affected by its mass. The mass of molecules attached to the nanosized string may therefore be determined by plucking the string and determining the frequency at which it vibrates. While this by itself is

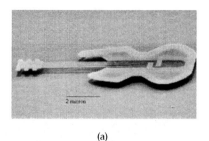

(a)

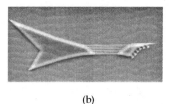

(b)

**FIGURE I.20**
Cornell University nanoguitars. (a) 1997 version; (b) 2003 version.

probably not the creation of "music" that would be protectable by copyright law, it does represent the very early stages of investigations of the acoustic properties of nanoscale devices. As this research continues, sequences of sound may well be produced that are meaningful and subject to the full protection that copyright law provides.

Copyright law developed mostly to provide a form of protection for the intellectual property embodied in creative works. This is one reason why some level of creativity remains a requirement to obtain copyright protection. This chapter has explored some of the ways in which nanotechnology research embraces the creative aspects that copyright law covers. In many ways, the examples that have been discussed remain as curiosities—the Osaka bull and the Cornell guitars are fascinating because they are so unusual. But there are many other nanotech structures that are now routinely being created that have creative aspects. They may not be as visually dramatic as the Osaka bull or the Cornell guitars, but in some ways many of them may provide even more important contributions to the way we view the world. It is copyright law that permits these creative endeavors to be protected. As Lawrence Lessig has acutely observed, "Of all the creative work produced by humans anywhere, a tiny fraction has continuing commercial value. For that tiny fraction, the copyright is a crucially important legal device."

### Discussion

1. *"Moral rights"* are sometimes considered to be a form of copyright and sometimes analyzed under other doctrines. Such rights attempt to recognize that the creators of works have a certain relationship with their creations, even after they have left the creator's ownership or possession. Changes to a work that would impact this relationship, such as in the form of distortions or mutilations, may violate the creator's moral rights. Do you think such moral rights have a place in the development of nanotechnology? What criticisms can you see being leveled at your answer?

2. In discussing the protection of the works, the main text asserts uncategorically that "there is no question that nanotech creations are never used as buildings." Do you agree? Is it possible that a *"building"* might be defined broadly as "something built or constructed" instead of a "structure for human habitation"—so that nanotech creations could then be considered "buildings" whose designs would be protected as architectural works? What if the nanotech creations housed other, smaller structures? Do you think such a broad meaning is what legislators had in mind when they provided the definition of "architectural work"?[41] What is your view of attorneys who try to extend the meanings of statutes in this kind of way?

## C.  Integrated Circuit Topographies

One of the major benefits of both patent and copyright protection is the breadth of subject matter that can be covered. Utility patents can be used to provide protection for an enormous variety of inventions, and copyrights can similarly be used to protect almost any physical manifestation of creativity. This breadth is available because of how these forms of intellectual property are defined—they are defined in relatively abstract terms, focusing on such features as utility, novelty, originality, creativity, and so forth.

There are other forms of intellectual property that are defined in more concrete terms, which makes their application much more restricted. One such example is the protection afforded to integrated circuit topographies. The basic idea of such doctrines is specifically to provide protection for the structures of integrated circuits—the way in which the active elements are laid out on a chip. The protectable structures are often referred to as *"mask works"* because the layout on the chip corresponds to the structure of a photolithographic mask that is used during the fabrication process. Such mask works may correspond to two-dimensional or three-dimensional layouts depending on the specific structure of the chip.

Intellectual-property protection for mask works arose to try to fill a gap between the protections offered by copyright law and the protections offered by patent law. A specific layout is not clearly within the realm of patentable subject matter, but because the mask geometry has a functional nature, it is also not clearly protectable under copyright law. There may be aspects of mask works that are protectable under either doctrine, but the application is not a tidy one.

This gap-filling approach is reflected in the rights that are provided in mask works. Similar to copyright, only the owner of a mask work (or someone authorized by the owner) may reproduce the mask work. And similar to patent rights, only the owner (or someone she authorizes) may import or distribute a semiconductor chip product in which the mask work is embodied.[42] One thing that is notable is that specific exceptions are provided to permit others to reverse engineer the mask work.[43] This promotes a public policy in which others are encouraged to seek improvements by designing around protected intellectual property. But at the same time, no generalized permission to engage in "fair use" of a mask work is provided, making it somewhat stronger than copyright protection in this respect.

In the United States, mask-work rights are defined by the Semiconductor Chip Protection Act. The title of this act already indicates one of the restrictions on the right that may impact how it can be extended to cover certain types of nanotechnology. Specifically, protection is provided only for semiconductor chips, with the act defining a *"semiconductor chip product"* as the final or intermediate form of any product "having two or more layers of metallic, insulating, or semiconductor material, deposited or otherwise placed on, or etched away or otherwise removed from, a piece of *semiconductor material* in

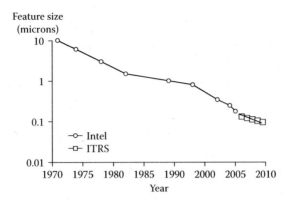

**FIGURE I.21**

A plot illustrating the systematic decrease in feature size for transistor components, shown for both Intel products and data reported by the International Roadmap for Semiconductors (ITRS).

accordance with a predetermined pattern."[44] In addition, the semiconductor chip product must be "intended to perform electronic circuitry functions."[45]

With these restrictions, there are certain nanotechnology products that clearly fall within the scope of the Semiconductor Chip Protection Act. As miniaturization of integrated circuits has continued, the feature sizes of the components have shrunk to the nanometer scale (see Figure I.21). It is becoming common to see reports of feature sizes in integrated circuits on the order of 100 nm, bringing integrated circuits themselves within the scope of nanotechnology. The term *"feature size "* is used in characterizing integrated circuits and defines the scale at which the smallest feature can be produced in a fabrication. The term is relatively generic since a "feature" could be a wire, a transistor, or some other component.

Application of mask-work statutes to integrated circuit technology, including at the nanometer scale, is straightforward because those statutes were originally conceived with integrated circuits in mind. While the nanometer scale is being reached in conventional integrated circuits, some have speculated that further decreases in size will be more difficult to achieve. The concern is that feature sizes are now about the same order in magnitude as the wavelength of light used in photolithographic processes so that further attempts to reduce the feature size will encounter limitations dictated by the laws of physics.

The answers proposed for addressing these limitations inevitably turn to the structures that are the bread and butter of much nanotechnology research: nanotubes. One proposal is to develop electronic devices that make use of the conductive properties of carbon nanotubes. One way of affecting these conductive properties is to introduce a twist into the nanotube so that it has semiconducting properties instead of purely conducting properties (Figure I.22).

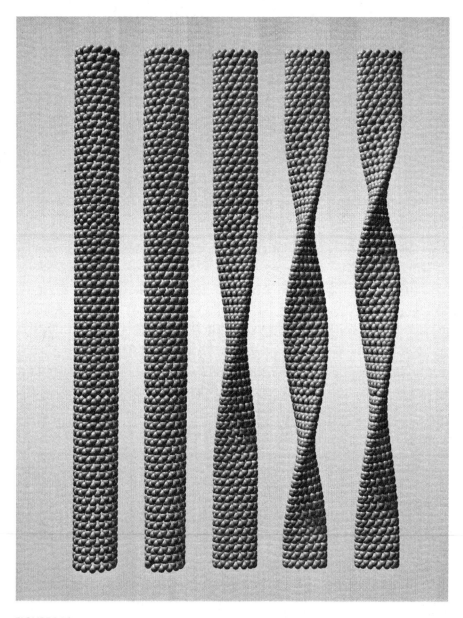

**FIGURE I.22**
The electrical-conductivity properties of carbon nanotubes may be determined by their degree
of twist. (Image reproduced with permisison.)

Appropriately, semiconducting nanotube structures may then be used in
the fabrication of alternative designs for transistors that may achieve even
greater reductions in feature size (Figure I.23). Because these kinds of structures may be fabricated from a piece of semiconductor material and perform

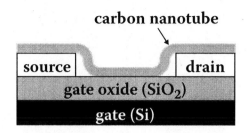

**carbon nanotube**

source | drain

gate oxide (SiO$_2$)

gate (Si)

**FIGURE I.23**

A proposed structure for a nanotube field-effect transistor that uses a semiconducting carbon nanotube to govern current transport between source and drain electrodes.

electronic-circuitry functions, they remain clear examples of structures that may benefit from protection under the Semiconductor Chip Protection Act.

It is interesting to consider further extensions of this pattern of development. For instance, a number of similar techniques are also being proposed to be applied to nanoelectromechanical structures (NEMS), which generally include mechanical aspects in the form of moving parts. The way this technology has been evolving blurs the distinction between complex mechanical systems and integrated circuit electronics. The batch fabrication techniques used in integrated circuit manufacture may be adapted to NEMS processing in producing such complex structures as sensors and actuators— the structures that have long been the most expensive and yet least reliable of many larger-scale electronics systems.

What is interesting about these kinds of NEMS structures, at least from the perspective of the Semiconductor Chip Protection Act, is the fact that they do not have purely electronic functionality; they also have a significant mechanical aspect. But this is unlikely to have any meaningful impact on the application of the act. It is still the case that they are intended to perform electronic functions and, as long as they are fabricated using silicon-based technology, will be made from a piece of semiconductor material.

It should be emphasized that these are relatively isolated examples of a wide range of nanotechnology structures that may be afforded protection as mask works. There are other nanotech structures where the application of mask-work protection is considerably less clear. Consider, for example, microfluidics devices (see Figure I.24). These devices are used in applications involving very small amounts of fluid, usually on the order of nanoliters. Fluids exhibit decidedly nonintuitive behaviors at such small volumes that are useful in certain applications, most notably a variety of biological applications.

For instance, one class of microfluidics devices are used as DNA chips to investigate how different genetic assays respond to different samples. The goal in these kinds of experiments is to check every possible assay–sample combination, an imposing task whenever there are more than a minimal number of assays and samples to be checked. Microfluidics devices provide an efficient and cost-effective way of completing this task by providing fluid pathways for tiny fluid amounts to check all possible combinations.

**FIGURE I.24**
A microfluidics chip produced by Fluidigm Corp. The chip, which is about 3 inches × 4 inches, allows researchers to investigate all combinations of one set of 96 fluids with another set of 96 fluids—a total of 9216 experiments conducted simultaneously on a single chip. (Image reproduced with permission of Flur Corporation.)

In many respects, such microfluidics devices are like integrated circuits. Indeed, they are sometimes described as *"integrated fluidic circuits."* An analogy is made between the electronic functions of conventional integrated circuits and the biochemical functions performed with integrated fluidic circuits. Just as an electronic integrated circuit is designed with transistors to control the flow of current through the device, an integrated fluidic circuit is designed with pumps and valves that regulate the flow of fluid. The analogy is sometimes extended to note that the miniaturization that has been characteristic of the development of electronic integrated circuits is being repeated in integrated fluidic circuits, which have progressively been made to use smaller and smaller volumes of material.

Even the fabrication techniques used for electronic integrated circuits have been borrowed in the development of integrated fluidic circuits. For instance, one set of techniques for producing microfluidics chips forms a series of layers over a substrate, with patterns being created in the different layers. Just like with integrated circuits, the patterns may be created by using a mask to expose different portions of a given layer to light and thereby have that portion respond differently. This type of photolithography is virtually identical to the sort of photolithography used in producing integrated circuits. It is this type of patterning through the use of a mask that gave rise to the terminology *"mask work"* when referring to electronic structures; it seems only natural to apply the same terminology when the same kind of patterning is used to produce fluidics structures.

While the analogy between fluidics and electronics is a compelling one, there remain certain difficulties in being able to apply the Semiconductor Chip Protection Act to microfluidics devices in the United States. The first part of the definition of a *"semiconductor chip product"* in the act is satisfied in many microfluidics devices because the substrate that is used is a semiconductor. But there is not nearly as strong a need to use a semiconductor substrate in fluidics applications—after all, it is the electronic properties of a semi*conductor* that define it is as such, and these electronic properties are not of much relevance in fluidics applications. There are, for instance, many microfluidics devices that are formed over other types of substrates and that would therefore seem to be outside the scope of the act.

Perhaps more important, though, is the fact that microfluidics devices rarely satisfy the second part of the definition of a *"semiconductor chip product"* since they are usually not "intended to perform electronic circuitry functions." Some specialized applications can be imagined in which electronic functionality is incorporated into the microfluidics device, but in the absence of such functionality, it seems that these kinds of devices will not be afforded the sort of protection that the Semiconductor Chip Protection Act provides. When evaluating the applicability of this type of protection to other kinds of nanotechnology devices, a similar analysis should be performed.

This particular impediment to applying the Semiconductor Chip Protection Act for certain types of nanotechnology also exists in the parallel laws of other countries. While not as nearly universal as patent or copyright protection, many countries do include provisions to protect integrated circuit topographies. In defining the products to which the topography applies, most other countries do not require that the base material be a semiconductor, but do require the performance of "electronic functions" in the final product.[46]

It is interesting to ponder, from the perspective of how intellectual-property rights develop, why this requirement of "electronic functions" exists. After all, the other types of intellectual property that were discussed previously are defined in much more abstract terms. It is true that patents and copyrights both technically have subject matter constraints; but when these constraints are examined closely, there is not much that is excluded—provided that some invention is useful in the case of patents or that there is something creative produced in the case of copyrights. Applying the same kind of logic that has traditionally been applied to these other types of intellectual property, it would seem that the more important aspect of mask-work protection would be the pattern of the mask work itself, without particular concern to the functionality of devices produced using it.

Earlier, I touched on the fact that mask-work protection attempts to fill a gap between the protection patents provide and the protection that copyrights provide. In the late 1970s, this gap seemed especially acute. There was a concern among manufacturers of semiconductor chips that it was too easy to duplicate a chip design that had potentially taken months or years of time and millions of dollars to develop. Essentially all that would need to be done was to take a chip that had been released, strip away its layers, and then use the chip as a model for large-scale copying of the structure. But the way in which mask works integrate functional aspects made it difficult to apply copyright principles to the designs, and the designs themselves were too specific to make patent protection of significant value.

It was in this context that the United States Congress first considered how to provide intellectual-property protection to semiconductor chips. The goal was to address what was viewed as a relatively narrow problem without disturbing other forms of intellectual property. At the time, other forms of integrated circuits were as yet unconceived so that their deliberations were focused centrally around electronic chips. This focus is manifested in the form that the act finally took and which served as a model for similar intellectual-property rights in other countries.

Although integrated circuit topographies are now protected in a number of countries of the world, this form of intellectual property has received very little practical application in the quarter century or so that it has been available. In some ways, this is surprising given the emphasis that was placed on the need for such protection in the early 1980s. The major reasons for this are the relatively short term that is provided for such rights and the need to register the mask works. Basically, the term of protection begins on the earlier of the date that a mask work is registered with the copyright office or commercially exploited. The term then nominally lasts for ten years from that earlier date. But if the mask work is not registered within two years of commercial exploitation, mask-work rights are lost.

Could the development of nanotechnology result in a revival of mask-work protection? The answer to this is as yet unclear. Certainly the possibility of such a revival would be greater if the requirement of "electronic functions" in the final product were relaxed. This could potentially happen in two ways. The most decisive route would be for parts of the nanotechnology industry to lobby governments to note that the development of technology in the last twenty-five years has exposed a weakness in mask-work protection that should be repaired legislatively. This kind of development would be consistent with a respectable history. As technology advances, weaknesses in the definitions of intellectual property have often been identified and the general trend is for legislation to respond by expanding the subject matter that can be protected rather than by restricting it.

Another possibility is for a court to examine the legislative history of the Semiconductor Chip Protection Act and to determine that integrated circuit topography protections should apply to mask-work structures that are used in producing even devices that do not perform electronic functions. After all, courts have been instrumental in the past in clarifying that certain types of technology are indeed entitled to intellectual-property protection under existing doctrines. Perhaps the most striking examples of this in recent years are judicial decisions in the United States that genetically modified organisms[47] and business methods[48] are protectable.

These kinds of interpretive issues are interesting ones and often engender debate about what the relative roles of the judiciary and legislature should be. When courts engage in this kind of interpretive exercise, some complain that the judiciary is overstepping the boundaries set for it by interpreting a phrase in a way that the legislature could never have intended. Others respond that the very fact that there was no possibility for the legislature to conceive of the issue means that courts with a modern perspective must step in and provide a modern interpretation.

The question always boils down to whether this kind of interpretation is eviscerating the intent of the legislature or giving effect to it. While arguments are made on both sides in all areas of the law, the development of new technologies provides an acute illustration of the interpretation difficulties that courts often face. Nanotechnology was in its infancy at the time the Semiconductor Chip Protection Act was passed and the notion that there could be integrated fluidic circuits as well as integrated electronic circuits was completely unknown. Does inclusion of the word *"electronic"* in the statute do nothing more than reflect the unavoidable ignorance of the legislature or were there good independent reasons to be limiting in this way?

While this example of judicial interpretation is raised here in the context of intellectual-property issues, it is worthwhile considering the difficulties that a new technology like nanotechnology presents for judges in all areas of the law. It may be useful to keep this issue in mind when considering the regulatory and liability issues discussed in other parts of this book.

### Discussion

1. Figure I.25 is an example of *"chip graffiti."* The photolithographic processes that are used to build up integrated circuits may be used in the creation of a variety of artistic figures that chip designers sometimes include on chip space that would otherwise be unused. What types of intellectual-property protection can you see being asserted for chip graffiti? Can you think of any practical reasons why it might be beneficial to include chip graffiti on chips? How do those reasons apply to nanotechnology products?

2. The main text suggests that the definition of a *"semiconductor chip product"* as "intended to perform electronic circuitry functions" may limit the application of mask-work protection to certain types of nanotechnology. Do you agree? Can you think of any way of circumventing this definition to argue that mask-work protection of layouts for microfluidics devices is protectable? One analysis might begin by considering whether it is always possible to discern from a mask work (by itself) whether a fully completed chip fabricated using the mask work will perform "electronic circuitry functions."

3. The United States Constitution specifically names the president to "be Commander in Chief of the Army and Navy of the United States."[49] Who is the commander in chief of the Air Force? Why? What interpretive principles did you apply in coming up with your answer? If you apply those principles to the question, "Is the patterned layout of a microfluidics device a mask work?" what answer do you give? Do you think these two questions are properly analogized? Why or why not?

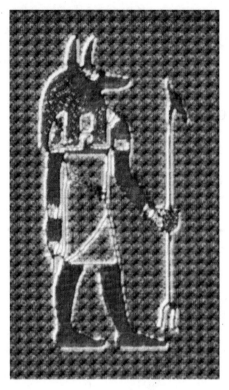

**FIGURE I.25**
A representation of the Egyptian god Anubis exists on the Silicon Graphics MIPS R12000 microprocessor. This representation has a height of about 100 µm. It is an example of chip graffiti. (Image reproduced with permission of Michael W. Davidson, Florida State University.)

## D.   Trade Secrets

You find yourself thirsty in New Delhi in the year 1985 and would like to purchase a bottle of your favorite soft drink, Coca-Cola®. Sorry, you can't. The reason is that between the years 1977 and 1991, the Indian government had foreign-investment laws that would have required the disclosure of trade secrets. In particular, for Coca-Cola to have been sold in India during that time would have required that the recipe for making it have been disclosed. When these laws came into effect in 1977, Coca-Cola, Inc., felt so strongly that it did not want to reveal its secret that it opted instead to shut down twenty-two established bottling plants in India and forgo selling its products in one of the most populous countries of the world. It was only once the Indian government relented in its position that sales of the popular drink in that country resumed.

Ever since John Pemberton, a pharmacist in Atlanta, developed the concoction known as Coca-Cola in 1886, the precise ingredients and their proportions have remained one of the most closely guarded industrial secrets. While Coca-Cola is known to have the same basic ingredients as every other cola—water, sugar, caramel, and caffeine—it also contains a secret ingredient identified to the world only as "Merchandise 7X." The actual composition of Merchandise 7X is reputed to be known only to a handful of individuals who are forbidden from traveling together lest a collective accident cause the secret to be forever lost. In fact, this restriction (should it actually exist) may be excessive since the composition is also supposedly stored in a vault in the Sun Trust Bank Building in Atlanta where it is protected under lock and key and with a guard posted twenty-four hours a day.

Up to this point, the different forms of intellectual property that have been discussed have a common factor: they all protect something that has been disclosed to the public. There are a wide variety of different circumstances in which it is more strategic to keep the intellectual property secret. And the decision to do so is one that is recognized by the law as a legitimate one. While the other forms of intellectual property include inducements provided by the government to encourage disclosure, the fact that there is also a body of law protecting trade secrets reflects the fact that disclosure is almost never compulsory.

The Coca-Cola example provides a vivid illustration of one of the most important consequences of deciding to keep intellectual property secret: trade secrets are not subject to any time limitations in the way that patents and trademarks are. If John Pemberton had decided in 1886 to patent Merchandise 7X, his patent would have expired sometime around the start of World War I. Since details of the composition would have been required in the patent, anyone could freely have consulted it and made his or her own Coca-Cola without infringing on any intellectual property for almost the last hundred years.

The Coca-Cola example may also illustrate the most significant drawback of trade secrets, which is that they provide no protection for independent discovery. How close in composition are competitor cola drinks—Pepsi®, Royal Crown®, Afri-Cola®, and so forth—to Coca-Cola? While this might be an important question in the marketing of these various colas, it is essentially an irrelevant question from the perspective of intellectual property.[50] The major risk of deciding to maintain intellectual property in the form of a trade secret is that it provides no protection from independent discovery. This risk of independent discovery is ultimately what underlies the effectiveness of the monopoly inducement provided by the patent system. If someone owns a patent, she is permitted to exercise the exclusion rights it provides even against independent later discoverers of the technology. This is not true at all with trade secrets, which are forever vulnerable to the possibility of independent discovery.

## 1.   What is a Trade Secret?

Like many legal concepts, the precise definition of *"trade secret"* is one that tends to have ill-defined borders. As more than one federal appellate court has observed, a secret "is one of the most elusive and difficult concepts in the law to define."[51] Different jurisdictions approach the question differently and have a variety of nuances in the language that they use. At its core, though, there are generally three requirements for some piece of information to be considered a trade secret: (1) it must not be generally known, at least to the relevant portion of the public; (2) its secrecy must provide its owner with some economic benefit; and (3) there must be reasonable efforts made to maintain its secrecy. Within this definition, there are a host of different kinds of information that have been recognized as being trade secrets.

Many of the categories of trade secrets are relevant to any type of company or organization, including nanotechnology companies. Such information as the identities of customers and customer preferences, the identities of vendors, marketing strategies, and company finances is information that almost no one would consider making publicly available unless subject to some legal requirement to do so.[52] These kinds of information are also the sort for which other forms of intellectual property offer no meaningful protection so the decision to maintain them as trade secrets is an easy one.

Other categories of trade secrets have more particular relevance to nanotechnology. These include information about certain kinds of manufacturing processes, research data and analysis, and testing protocols. In many instances, the decision whether to maintain this type of information as a trade secret is less clear. For example, a legitimate question may be raised about whether to keep a certain manufacturing process secret, thereby running the risk of its independent discovery by a competitor. The alternative is to attempt to patent the process, which has the advantage of providing a mechanism for preventing competitors from using it. Similar considerations may apply to testing protocols, although in most cases the kinds of factors that weigh into the decision argue more in favor of keeping such protocols as trade secrets. In purely commercial contexts, it is often also advisable to keep underlying research results secret, although other factors frequently prompt disclosure of at least some of that information. Those who develop nanotechnology in academic settings are particularly subject to pressures to disclose results since they are often evaluated on their publication records, and both academics and researchers in small companies are frequently supported by government funds that include disclosure requirements.

While these kinds of information have the potential to be identified as trade secrets—and thereby to enjoy the protections that the law provides to trade secrets—any of the three factors listed earlier has the potential to disqualify the information. This is perhaps best understood with a specific example. Consider a nanotechnology company, NanoD, that produces dendrimers for drug-delivery applications. A dendrimer is a type of polymer molecule that has a branched structure like that shown in Figure I.26. From

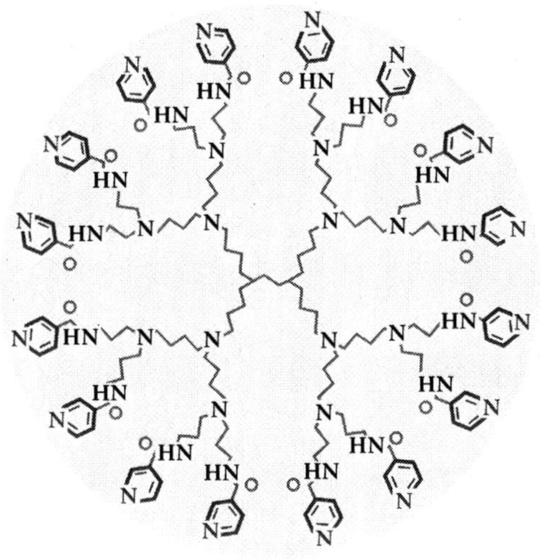

**FIGURE I.26**
A schematic illustration of the structure of a dendrimer.

a core functional molecule, there are successive "generations" of molecules built up in branches, producing a structure that resembles a tree (or any of a variety of other fractal-type structures). Proposals for using dendrimers in drug-delivery applications arise from the fact that they can be fabricated to be "artificial cells"—they have a relatively hollow core surrounded by a compact surface and may be synthesized to have a total size similar to that of cells. The surface acts to encapsulate the drug until it can be delivered very precisely to the area of a body where it is to be applied.

In this example, scientists at NanoD have developed a synthesis technique that greatly reduces the cost of producing their particular dendrimer structure. Recognizing the potential for this technique to provide it with a competitive advantage, NanoD decides not to disclose it in any way—they do not file for a patent application and they prevent the scientists from publishing papers describing it or talking about it at conferences.

But suppose that the scientists were, in fact, unaware of some relevant research by others that disclosed the same technique. In that case, the technique would not qualify as a trade secret no matter what actions NanoD took because of the first factor mentioned above, that is, the information was generally known to the relevant portion of the public. In applying this factor, the actual knowledge of the company and its scientists is irrelevant. All that matters is that the information was already known. This is not a surprising result—a state of ignorance on the part of those who develop information should not be the basis for generating intellectual-property rights.

The converse to this scenario is where the NanoD scientists were the first to develop the synthesis technique, but it was later disclosed by someone who learned it from them. When such a disclosure is made improperly, it will not necessarily defeat the status of the information as a trade secret. This is particularly true if there are ways of minimizing the extent of the disclosure so that it never becomes readily accessible to the public at large.

The second factor is, in some ways, more interesting. It is important to recognize that this factor requires that the secrecy of the information provides some economic benefit. It is not enough that the information itself provide such a benefit. For instance, in this case, it is apparent that the efficiency that the technique provides permits costs to be reduced for NanoD and that it thereby enjoys some economic benefit. But does the secrecy of the technique also provide an economic benefit? While the answer to this question is almost certainly yes, it is not as immediately apparent that it must be so. It is possible to imagine scenarios that at least lend themselves to arguments that the secrecy is not providing an economic benefit. Suppose, for instance, that no other company uses the same dendrimer structure because NanoD holds a patent on it, and the technique is of value only in producing the dendrimer structure used by NanoD. In such a scenario, is the secrecy of the technique providing any economic benefit to NanoD or is it merely the application of the technique itself that provides all the economic benefit?[53]

The third factor is undoubtedly the most interesting. It is not enough merely to be in possession of some valuable secret information to enjoy the benefits of trade-secret laws. There must, in addition, be some active efforts to maintain the secrecy of the information. An important part of the activities of virtually every nanotechnology company should include implementing a trade-secret program to enjoy the benefits that trade-secret protection provides.

The level of effort should be approximately commensurate with the importance of the secret. In most cases, it is not necessary to lock the only copy of the information in a vault that is watched by a full-time guard. But, at a minimum, copies of the secret information should be identified as confidential and access to the secret information should be limited to those in the company who have a legitimate need to know it. Restricting access to information may be performed physically, such as by having areas that only some employees are permitted to enter. Access may also be restricted technologically by using software protections to limit those who have electronic access to certain information.

Other actions that should be taken with employees of the company include programs to educate them about trade secrets. They should understand the need to mark materials as confidential, have nondisclosure agreements executed when there is a reason to be discussing a trade secret with someone outside the company, and so on. It is preferable that employees sign an employment agreement in which they acknowledge the value of the company's trade secrets and agree to keep them confidential. When employees leave the company, an exit interview should be conducted in which they are reminded that their obligations to maintain the confidentiality of the company's secrets continue even after their employment ends.

It is also best if security mechanisms are in place at the company so that visitors will not improperly have access to secret information. They should be required to sign in and out when they visit and should have an escort whenever they are in areas where trade-secret information might be available. If visitors are to be shown trade-secret information, they should be warned that the information is confidential and asked to sign nondisclosure agreements in which they agree to keep it confidential.

Of course, this is not an exhaustive list of the steps that need to be taken to preserve the confidentiality of trade secrets, but it does give a flavor of the kinds of activities that courts will look to in evaluating whether the third requirement has been met. There is, thus, some effort involved in maintaining trade secrets and a decision inevitably needs to be made whether some piece of information should be protected as a trade secret or protected using some other form of intellectual property. Almost always, this decision amounts to deciding whether to apply for a patent and disclose the information in the patent application or to keep the information as a trade secret.

## 2. Patent or Trade Secret?

There are a host of considerations that apply in making such a decision. Perhaps the most fundamental consideration is simply cost. No matter what the size of an organization, its intellectual-property program will be constrained by a budget. It is virtually always the case that the cost of maintaining some information as a trade secret will be less costly than obtaining a patent for it—the process of applying for a patent requires the payment of government filing costs and attorney's fees, which can be significant. The inventive productivity of an organization, particularly of the typical nanotechnology organization, routinely generates so many inventions that they cannot all be protected by patents. The reality is that organizations who are addressing their intellectual property in a serious way are not faced with the question of whether to keep trade secrets, but must decide which of the various inventions should be patented and which should be kept secret.

Making such a decision should take into consideration the relative ease of reverse-engineering the technology. If it is relatively easy to reverse-engineer, the technology is a better candidate for patent protection than for trade-secret protection. This is because even the most stringent of efforts to avoid direct disclosure of the information will not prevent a competitor from simply purchasing a product and analyzing it to uncover the "secret." Because of this, fabrication processes are often good candidates for trade-secret protection because it may be impossible to tell from the final product what particular processes were used in creating it.

The converse of this is accounting for the likelihood of independent discovery. If the technology is the result of a pattern of research that is likely to be duplicated by others, it is again not a good candidate for trade-secret protection. While the patent laws provide protection against independent discovery, giving the superior rights to those who first make an invention, there is no such protection for trade secrets.

Another factor that may play a role in the decision are whether there is any realistic likelihood that protection will be needed longer than the twenty-year term of a patent. In the case of nanotechnology, it is not clear what the answer to this might be. Technology that has biological applications might well endure for more than twenty years, but nonbiological applications are likely to have a shorter lifetime. It is also useful to evaluate whether there might be any prior art or other issues that would cause difficulties in obtaining a patent. The most clear-cut example of this is where the information embraces something that is simply not legally patentable, leaving trade secrets as providing the only viable form of intellectual-property protection.

A final consideration that should sometimes be made is to account for the ultimate disposition of the intellectual property. Many start-up companies, for example, are created with the expectation that they will be sold within a certain, relatively short, period of time. Under such a business scenario, a more aggressive patent position is usually advantageous, making the company more attractive to potential purchasers. In a similar way, any technology that has the potential to generate significant licensing revenue as part of a licensing program may be better protected with a patent. Licensing of technology with patents is relatively well defined while attempting to do something similar with trade secrets is fraught with a variety of pitfalls.

## 3.  Remedies

If all the requirements have been met to establish some piece of information with the status of a trade secret, its improper disclosure has the potential to be addressed in either a civil action or a criminal action. While nanotechnology companies will mostly be concerned with civil actions since they provide a mechanism for them to recover if their trade secrets are disclosed, those who have access to nanotechnology trade secrets would be well advised to be aware of the potential for criminal liability. Nanotechnology is one of those technologies that is sensitive enough to implicate the criminal statutes under certain circumstances.

The kinds of remedies that the civil statutes identify are those that are rather typical of civil statutes—generally damages, injunctions, and attorney's fees. Essentially what this means is that someone who improperly discloses a trade secret can be required to pay an amount of money that represents the loss suffered by the owner of the trade secret. In many statutes, these monetary damages may be augmented in the case of "willful or malicious" misappropriation of the trade secret, with the Uniform Trade Secrets Act providing for as much as treble damages in such circumstances. It is also possible for a court to order that some action be taken in the form of an injunction, usually in the form of an order to stop any ongoing disclosure that might be taking place of the trade secret—for instance, removing the information from a Web site where it is disclosed.

In the United States, criminal penalties for trade-secret misappropriation are provided by the Economic Espionage Act, which was enacted in October

1996. The penalties that exist under the act are entirely punitive. They are the most severe punishments available for violation of an intellectual-property right. A person who improperly discloses a trade secret under this act can face up to ten years imprisonment and be fined up to $500,000. If the offense is committed knowingly to benefit a foreign power, the term of imprisonment can be as lengthy as fifteen years. If the violation is committed by a corporation, it can be fined up to $5 million, with that ceiling being increased to $10 million when it is done to benefit a foreign power.

The severity of these punishments is a reflection of the value that trade secrets can have and of the importance with which they are now viewed. Before the passage of the Economic Espionage Act, it was difficult to formulate a criminal theory under which someone could be prosecuted for theft of a trade secret. And it was essentially unheard of for someone to be imprisoned for such theft. But since then, a number of prosecutions have taken place under the act and people have been incarcerated as a result of those prosecutions. The first case in which a conviction was achieved under the act occurred in December 2006 when Fei Ye and Ming Zhong pleaded guilty to two counts of economic espionage in the U.S. District Court for the Northern District of California.

In an action that seems to be more the stuff of cinematic fiction, the pair was arrested at San Francisco International Airport with stolen trade secrets—blueprints and computer-aided design scripts—stuffed in their luggage. They had plans to develop a microprocessor for their company, Supervision, Inc., and had stolen trade secrets from Sun Microsystems, Inc., and Transmeta Corp. that they were planning to use in their technical development. An arrangement with the Chinese government for funding of Supervision would have resulted in their sharing profits from the resulting chip sales with the City of Hangzhou and the Province of Zhejiang in China—an episode of economic espionage that implicated parts of a foreign government.

Nanotechnology is an area that is ripe for trade-secret litigation. It is a technology that is developing quickly and in which there is significant movement of researchers among different organizations. This provides an environment in which a variety of different people have access to trade-secret information and in which there is a strong incentive to attempt to gain an advantage over competitors. This combination results in an atmosphere where the temptation to make improper use of trade secrets is strong.

Indeed, despite the relative infancy of nanotechnology, there have already been a number of trade-secret actions that have received media attention. Perhaps the most instructive of these concerns Dr. Donald Montgomery, who was employed as a senior research scientist by Nanogen, Inc., between May 1994 and August 1995. Nanogen is a nanotechnology company that develops and markets microarrays used in DNA analyses. In 1996, Montgomery founded CombiMatrix Corp., a nanotechnology company that developed biochips. In January 1998 and September, 1999, Montgomery filed patent applications directed to certain aspects of this biochip technology and assigned those applications to his company.[54]

Nanogen sued. In November 2000, it filed a lawsuit alleging that Montgomery had misappropriated the technology described in those patent applications, that is, that he had stolen and disclosed trade secrets owned by Nanogen. After the lawsuit progressed for some time, the parties settled. The terms of the settlement included a payment to Nanogen of $1 million and 17.5 percent of the CombiMatrix stock (which by this point had become acquired as a unit of Acacia Research Corp.). The total value of this settlement was estimated to be about $11 million.[55]

Whether the allegations against Montgomery were accurate and whether he had any legitimate defenses are not relevant to the discussion here. Instead, what is significant is the fact that the circumstances under which these allegations took place are common in nanotechnology. As a relatively nascent industry, there are strong incentives for scientists and engineers to wish to form their own start-up companies to explore and develop their own ideas fully. And it is often difficult to delineate when that development merely exploits the skills that the individuals have acquired in their past employment and when it crosses the line into misappropriation of a trade secret.

These difficulties are, of course, exacerbated because the different parties inevitably have very different views of what has happened. When an employee leaves a company to form a competitive start-up, she is likely to be viewed by her former employer as a defector—a person who was mercenary in acquiring knowledge and skill from her employer only in preparation for a betrayal. The employee is instead likely to view herself as an industrious employee who contributed diligently to her employer but is now ready to take the next step in her development. The former employer does not want the skills his employee learned from him to be used against him, and the employee thinks she cannot remain indentured to her employer forever.

There is no question that these are difficult issues, both for scientists who work in nanotechnology companies and for those who employ them. All parties are well advised to be cautious in how they treat trade-secret information. The stakes are high. And in many ways this form of intellectual property provides the greatest opportunities for costly missteps.

### *Discussion*

1. Suppose Company X develops a technique for efficiently growing very long nanotubes and decides to keep its technique secret. The steps it takes to maintain this information as a trade secret are all reasonable and fully comport with every legal requirement. Several years later, when Company X is extraordinarily successful because of this technology, you independently discover the technique. If you apply for a patent for the technique (and assuming no other prior art or other issues), will it be granted? If so, can you use the patent to prevent Company X from continuing to use its technique? Do

you think your answers to these questions reflect a sensible public policy? What criticisms can you level at your answers from the perspective of public policy?

2. With similar facts to those in question 1, suppose that when you discover the technique for growing long nanotubes, you are employed by Company Y. Unaware that Company X uses the technique, you and Company Y decide to keep the technique secret and take all steps to afford the technique the status of a trade secret. Who owns this trade secret? If person Z learns of the secret from Company Y and discloses it publicly, does Company Y have a cause of action against Z? What if Z is the president of Company X and makes the disclosure after learning that Company Y uses the same technique? Is your answer different if he only identifies that Company X uses the technique instead of identifying that Company Y also uses it?

3. What is the best form of intellectual property protection for each of the following and why? In which of the following cases do you need additional information to determine the best form of protection? What additional information do you need?

   a. A new NEMS sensor design

   b. A newly discovered physical law that operates at certain nanometer scales

   c. Results of a survey of veterinarians to assess practitioner views of treating bone fractures in horses using a new nanotechnology process

   d. A necklace pendant having a complete version of the Qu'ran inscribed with a nanotechnology etching process

   e. A new method for fabricating nanotech optical coatings to be used on car windshields

   f. A new method for delivering drugs with nanodevices to treat tumors in humans

   g. Discovery that a compound routinely used as a component in cosmetics can catalytically promote self-assembly of structures useful in delivering stem cells for postsurgical tissue-repair treatments

   h. Results of ceramic fracture tests to identify potential customers of a new nanotech ceramic fabrication process

   i. A novel methodology for generating visualizations of nanotech structures based on their quantum-mechanical interactions

   j. A computer program to generate images using the methodology of i

   k. An internet Web site advertising the methodology of i (Are your answers to i, j, and k consistent? If not, can you devise a strategy to best accommodate the combination?)

    l.  A method of using swarms of airborne nanorobots in performing espionage

   m.  Lightweight military armor that incorporates nanometer-sized devices

## E.  Ownership of Nanotech Intellectual Property

With the progressive miniaturization that has characterized semiconductor processing for decades, many semiconductor light-emitting diodes (LEDs) may now properly be considered the result of nanotechnology fabrication processes. These LEDs are ubiquitous. They are seen in almost every lighting application imaginable—as indicators on car dashboards, cameras, and cellular telephones, among others. They are also increasingly used in traffic signals instead of conventional incandescent bulbs—each of the red, yellow, and green lights is actually formed as an array of LEDs with the appropriate colors. The use of LEDs in these many applications is beneficial because their energy consumption is small even while their light output is large; they do not generate waste heat to nearly the degree of conventional incandescent forms of light and can save significant amounts of money.

Many who obey the "stop" and "go" commands of traffic lights are, however, obliviously unaware of the David-and-Goliath drama that they represent. For decades, it had been well known how to make LEDs that would produce red light. But it was notoriously difficult to fabricate LEDs that would produce green, blue, or violet light—namely, light with shorter wavelengths that could make data storage on optical disks more efficient, that could be used as a component in making white-light devices, and that could be used to tell drivers when to "go."

The fundamental development that allowed these short-wavelength LEDs to be produced in any commercially meaningful way was made by Shuji Nakamura when he was working at a small Japanese company known as Nichia Kagaku Kogyo. Almost all of the development on this was done by him after hours, on his own time, because management of the company was unwilling to provide any meaningful support for the research. Today, the word *breakthrough* is too often used in a cavalier fashion to describe even modest scientific developments, but this was an example where the term genuinely applies. After announcement of Nakamura's invention in 1990, the company grew extremely rapidly, from having about 400 employees to now having around 3500 employees, with some ¥200 billion (about $2 billion) in annual revenue. Nakamura's invention spawned huge investments by companies and academic researchers around the globe that have resulted in the tremendous number of consumer products that today incorporate the fruit of his discovery.

And what was Nakamura's reward? As a supplement to his regular salary, his company paid him a bonus amounting to ¥20,000 (about $150). Nakamura responded by claiming ownership of the Japanese patent where the technology was first disclosed, spawning litigation between him and Nichia that lasted for years. The litigation proceeded in somewhat of a seesaw fashion, reflecting the fact that both parties had some legitimate claim to ownership rights in the patent. In 2002, the Tokyo District Court ruled that the patent was owned by Nichia, largely due to the fact that they had paid him ¥20,000 in direct compensation for it. But the court also ruled that Nakamura was entitled to a portion of the proceeds from the invention and ordered Nichia to pay him ¥60.4 billion (about $190 million), half of Nichia's profits that were attributed to the sale of products based on the technology. A subsequent appeal to the Tokyo High Court was met by chiding from that court, encouraging the parties to settle. They eventually did, with Nakamura receiving a payment of ¥843 million (about $1.8 million) from Nichia in 2005.

This example serves as a compelling reminder of the significant value that may exist in intellectual property. While Nakamura was able to recover money despite a ruling that he did not actually own the patent, the ability to reap the benefit of intellectual property is most directly tied to ownership rights. In the case of some inventions, ownership is clear. This is true, for example, where an independent inventor develops a gadget in his garage and files a patent in his own name. But this is an unlikely scenario for the development of much nanotechnology.

## 1. The Nature of Property

Instead, most nanotechnology development is occurring with collaborations among multiple scientists and engineers in universities and companies, often with the support of funding provided by governments. In these kinds of scenarios, there are numerous parties who can legitimately claim that a particular invention would never have been developed without their contributions—each of the scientists and engineers in the collaboration may have contributed to the ideas that developed into the final product; the company or university may have contributed the tools and overhead that were necessary to develop the products, not to mention the salaries that they paid to finance the development; and governments may claim that the funds that they provided were instrumental in supporting the discoveries. How can ownership be allocated in these kinds of circumstances? What kinds of factors determine who will have the strongest claim, particularly if disputes arise among the parties?

The most fundamental rule of ownership for intellectual property is that ownership begins with the person who creates it. This is generally true irrespective of whether the intellectual property takes the form of an invention that results in a patent, is in the form of a creative work that results in a copyright, or has some other form. Once that intellectual property has been created, though, it is treated like almost any other form of property in terms of the ability to transfer ownership rights.

Property rights are often described as a *"bundle"* of rights. An effective analogy that relies on this description compares property rights with a bundle of sticks. Each of the sticks represents some particular right that can be separated from the bundle. Thus, for instance, an owner of a piece of land—"real property"—is able to grant a variety of different rights without transferring ownership of the land itself. She can grant logging rights that allow loggers to remove trees, she can grant mining rights to allow miners to remove minerals from the land, she can grant a leasehold interest that allows a tenant to take possession of the land, and so forth. Each of these rights can be thought of as a twig in the bundle that the landowner gives away in transferring a property right. But as long as the owner keeps the bundle, she remains the owner of the land.

Subject to a few regulations, the scope and nature of the various rights that can be granted are limited only by the imagination of the parties to the transaction. In some cases, the rights that are granted are transferable—the recipient can give the stick to someone else—while in other cases they are not. Some rights are of limited duration—the recipient must return the stick after a defined period of time—while others are perpetual. In many cases, some of these rights may be outstanding when the owner decides to transfer ownership of the property—when she hands over the bundle of sticks, there are still a few twigs outstanding so that the new owner is subject to some encumbrances on the property. One example of this could occur when a landlord sells an apartment building; each tenant's lease prohibits the new owner from evicting him as a trespasser until the terms of the lease have been satisfied (and that twig accordingly returned to the bundle).

These same principles apply to intellectual property. The owner of a patent or copyright can transfer certain limited rights to others without necessarily transferring ownership. Again, the scope and nature of the rights that can be transferred are wide. The right to use intellectual property is referred to as a *"license"* and can take the form of a right to make a product subject to a patent claim, perform a piece of music subject to a copyright, etc. These rights can be of limited duration or perpetual, they can be limited in geographical scope, they can be transferable or nontransferable, they can be exclusive or nonexclusive, and so on.

## 2.  Joint Ownership

The ability to transfer these various property rights becomes especially interesting when the ownership is *"joint,"* that is, shared among multiple people. Because ownership of intellectual property originates with its creators, joint-ownership situations are a common result of collaborations that produce inventions or creations. When multiple people create intellectual property together, each of them shares in all the rights and has the ability to transfer those rights.

This is profound. If a team of five people develop an invention and are granted a patent to it, any one of them can grant any of the intellectual-property rights it represents to another—even against the wishes of the others. If the inventors have different interests, no matter what the reason, each of their interests can be frustrated by the others. If a group of four of them were developing a product in anticipation of selling it, their ability to use the patent in trying to enhance their market share could be thwarted by the fifth freely granting licenses to the patent to any potential competitors.

Because of this, some effort is usually made to have the intellectual property owned by a single entity, very frequently a corporation. This ensures that decisions, whether they be good or bad ones, are at least made with some definiteness. When a corporation owns the intellectual property, decisions about how to use it—whether to grant licenses under it, whether to sue others to prevent infringement, and so forth—are made according to the mechanism that the corporation uses to make any of its decisions. Depending on the articles under which the entity is incorporated, such decisions could be made by a vote among directors or shareholders, or by administrative actions taken by officers of the corporation.

## 3. Assignment Obligations

There are a number of different ways that corporations can acquire ownership of intellectual property developed by their employees. The most common mechanism is a contractual one, usually embodied in an employment agreement that the employee is required to sign when he joins the company. In exchange for employment, the employee agrees that intellectual property developed during the course of that employment will be owned by the employer. Because the development of intellectual property is so central to much of what nanotechnology companies do, they are well advised to have effective employment agreements in place to ensure that their ownership of inventions and creations by their employees is clear.

A common question that arises among people subject to such employment agreements is "What if I invent something at home without any support from my employer?" This actually occurs quite frequently. Inventive people are, after all, just that. They tend to have inventive ideas in a variety of different areas, not just in the specific areas of interest to their employers. Like many legal questions, the answer depends on a number of different factors. Most employment agreements include language that provides an exception for intellectual property that is developed on the employee's own time and without support from the employer, either in tangible form provided as tools and equipment, or in intangible form as trade-secret information that the employee has learned because of his employment. Even if an employment agreement doesn't specifically call out such an exception, courts generally read the language in such agreements restrictively so that such an exception is very often inherent.

The result is that, in many cases, such "home inventions" are owned by the employee and not his employer. But there is a critically important class of people to whom this exception does not apply: those who are *"hired to invent."* Whether someone falls into this category is not clear-cut; it is a highly fact-specific standard that ultimately hinges on the ability of the employer to provide evidence that the employee was hired to solve a particular problem and that the invention at issue relates to that problem. If a person is found to have been "hired to invent," though, it no longer matters whether the invention was made on the employee's own time. The employer owns it. In fact, it does not even matter under these circumstances whether there is an employment agreement. If the employee was hired to invent, ownership will rest with the employer irrespective of how, where, or when the invention took place.

As a field that is still in the process of emerging, nanotechnology is such that a relatively large fraction of employees of nanotech companies have been hired to invent. The scientists and engineers employed by such companies know that they have been hired to solve certain problems and to advance the development of specific areas of nanotechnology. All their inventions in those areas are therefore likely to be owned by the companies that employ them, with the technicalities of when or where they developed those solutions being irrelevant. This is an eminently reasonable state of affairs. After all, the companies are paying people with particular expertise to solve certain problems, and are entitled to own the solutions.

## 4.   Government Interests

The considerations described in the previous section may be complicated even further when support for the research that leads to the development of the intellectual property is provided by a public source. This is a particular concern for nanotechnology, with a variety of governments currently providing funding mechanisms to support nanotechnology research. There are differing views regarding ownership of intellectual property, particularly patents, when it results from funds provided by the public through their governments.

One line of thought is that a government that supports research resulting in a patent should retain ownership of the patent, thereby allowing it to be placed in the public domain. There is a good deal of common-sense appeal to this view. Patents are effective mechanisms for concentrating market power. Permitting corporations to acquire ownership of them from research funded by the public looks to many like a giveaway of a public benefit to a private interest. Consumers who pay taxes to support the research then end up having to pay monopoly-level prices for the fruits of that research.

The opposing line of thought focuses on the real need for incentives in attracting qualified firms to perform research for the government. In many instances, particularly in an area expected to have significant commercial application like nanotechnology, there is a combination of private and public funding sources available. If governments insist on retaining ownership of patents from the research they fund, they will lose in the competition for the best researchers. Instead, the government will find itself perpetually in the position of having access only to second-rate research performed by those who lost the competition for private funding sources.

For decades, the first view was the predominant one in the United States. But in 1980, the United States passed the Bayh–Dole Act,[56] which attempted a compromise between these positions. The basic thrust of the Bayh–Dole Act is that recipients of federal funding may be able to retain ownership of inventions that result from that funding in exchange for commercializing the inventions. There are a number of peripheral responsibilities that the owner must undertake as part of the commercialization: it must file for patent protection must be active in promoting the invention, and must give preference to U.S. industry in small business as part of its commercialization efforts.

The government does retain certain significant interests. The owner of the invention is required to disclose the invention to the particular funding agency and is significantly constrained in its ability to assign rights to the technology. Any royalties that are generated must be shared with the inventor, and excess income must be used for education and research purposes. Perhaps most important is the fact that the U.S. government must be granted a license to the technology (a "nonexclusive, nontransferable, irrevocable, paid-up license to practice or have practiced on its behalf throughout the world").

By vesting ownership of the technology in the organization that actually developed it, most of the concerns about there being insufficient incentives to compete for the best scientists and engineers are addressed. And by ensuring that the government retains a license, most of the concerns about the public giving up all interest in the research it funds are addressed.

It is worth noting that the Bayh–Dole Act does include provisions for rather draconian action by the government in the form of *"march-in rights."* These permit the government to intercede and force the owner of the patent to grant a license to other parties (or to grant the license itself if the owner refuses). This power arises only when the owner has been failing in its duties of commercializing the invention or where there are certain health or safety needs. In actual practice, the government has never exercised march-in rights in the entire quarter-century period that the Bayh–Dole Act has been in force. Indeed, there have only been a handful of requests for it to do so, and they have so far all been denied.

The prevailing view is that the compromise that the Bayh–Dole Act represents has been tremendously effective. Before its passage, the U.S. government had only been granted 30,000 patents, with only some 5 percent of them having been commercially licensed. What the Bayh–Dole Act accomplished was to include a variety of recipients of federal funding, most notably universities, in the activities of commercializing technology. "Technology-transfer" departments of universities have since grown considerably in importance since passage of the Bayh–Dole Act. They provide a mechanism by which the universities can generate licensing revenue from research, while at the same time assisting the development of new start-up businesses that are important to a national economy. In 1980, universities were filing patent applications at

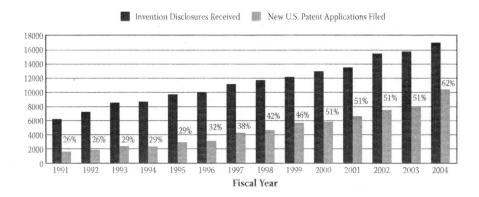

**FIGURE I.27**
Growth of invention disclosures and patent filings by universities after passage of the Bayh–Dole Act (Source: Association of University Technology Managers Licensing Survey: FY 2004).

the rate of a mere 250 a year; the data in Figure I.27 shows that this number has now grown to exceed 10,000 patent applications a year.

As an emerging technology, many of the patent applications that claim aspects of nanotechnology will be filed by universities through their technology-transfer offices. Development of the inventions that are disclosed and claimed will have been supported by any of a variety of government grants. In the United States, much of the funding for nanotechnology has been coordinated through the National Nanotechnology Initiative, an umbrella organization that cooperates with some twenty-five government agencies to harmonize efforts in promoting nanotechnology. Since its inception in 2001, over $6.5 billion has been invested in the organization.

This an impressive commitment by the public to support development of this technology. The funds that the public provides flow through the different government agencies as grants to university researchers, and the patent applications that result are owned by those universities. The universities in turn actively seek licensees in the form of start-up companies who will work to commercialize the technology. Many of these start-up companies will be formed by the professors and students who develop the technology at their universities, and who will be encouraged to contribute in this further way by the potential for further financial gain from their discoveries.

The experiences of the last twenty-five years have proven this to be an effective model in promoting the development of technology—and nanotechnology will benefit from the confidence that now exists in this model. While decades had previously passed where it seemed somehow unfair that private ownership of intellectual property could result from public funding, the benefits of doing so are now clear: there is more rapid development and commercialization of new and useful technologies, and this development generates significant employment opportunities. These are two of the most important contributions that any government can make.

It is true that, under this model, the government will retain many rights in the intellectual property that results, but this is unlikely to deter much of the commercialization efforts. The government has shown restraint in exercising its march-in rights, so much so that it would now seem anomalous if it ever did. While such rights may have seemed advisable when the Bayh–Dole Act was enacted, there has been a general change in viewpoint. It now seems apparent that the objectives of effective commercialization are readily being achieved with the incentives that ownership provides without the need for government intervention to move them along.

### Discussion

1. There are many occasions when inventions are developed jointly by employees of different companies. This can happen, for instance, when a supplier is attempting to provide a product that addresses a problem faced by a customer: the supplier's engineers and the customer's engineers jointly develop a patentable solution. Who owns such inventions? Would your answer be affected by the existence of applicable employment agreements signed by all the inventors agreeing to assign their inventions to their respective employers? What potential difficulties do you see with such circumstances? What advice would you give to each of the companies to avoid such difficulties? Is there anything the companies should do in advance of the collaboration that might help resolve the ownership issues?

2. Suppose a gifted scientific employee of a nanotechnology company conceives of an invention around the time that she decides to seek other employment. Preferring to make a good impression at her new job and having little motivation to help her current employer, she keeps her idea secret, only disclosing it a few weeks after beginning her new job. Who owns this invention? What additional information might you need to decide? Can you think of any provision in her employment contract that the original employer could have included to prevent this type of behavior? Do you see any public-policy objections to such a provision?

3. Sometimes inventions are developed by employees who are neither hired to invent nor have employment agreements that address ownership of the inventions. If such an employee develops an invention during the normal course of employment, that is, on company time and using employee resources, who owns the invention? Under the *"shop right"* doctrine, ownership is vested in the employee with the employer granted a royalty-free, nontransferable, nonterminable license to use the invention for its own purposes. Is this a fair apportionment of rights? Why or why not? Can you construct any examples where this seems unfair to the employer? What about examples where it seems unfair to the employee?

4. Under a copyright *"work for hire,"* a work created by an employee hired to make the work is owned by the employer and not by the employee. Compare this copyright doctrine with the "shop right" and "hired to invent" patent doctrines. In what ways does this comparison highlight differences in the rules of ownership for copyrights and patents? Can you provide any public-policy rationale that accounts for these differences?

5. There is a provision in the patent act that requires that assignments of interests in patents be recorded in the patent office,[57] in much the same way that mortgages and sales of real property are recorded with a county or city clerk. Why do you think such a requirement exists? What risks would there be to potential purchasers of patents without such a requirement?

---

## Notes

1. King, Ross. 2000. *Brunelleschi's Dome: How a Renaissance Genius Reinvented Architecture*. New York: Walker & Company.

2. Shirai, Yasuhiro, Osgood, Andrew J., Zhao, Yuming, Kelly, Kevin F., and Tour, James M. 2005. Directional Control in Thermally Driven Single-Molecule Nanocars. *Nano Letters* 5: 2330–2334. This article, describing the original nanocar, was the most accessed article of all American Chemical Society journals in 2005, reflecting the wide interest in general nanotech structures.

3. Morin, Jean-Francois, Shirai, Yasuhiro, and Tour, James M. 2006. En Route to a Motorized Nanocar. *Organic Letters* 8: 1713–1716.

4. 35 U.S.C. §112.

5. 35 U.S.C. §101.

6. *Diamond v. Chakrabarty*, 47 US 303 (1980).

7. 35 U.S.C. §102.

8. 35 U.S.C. §103.

9. *Interferences* are quasi-judicial processes that allow proof to be submitted to establish which of multiple parties claiming to have developed an invention did so first. 35 U.S.C. §135. The proceeding is usually presided over by a panel of three patent law judges who sit on the Board of Patent Appeals and Interferences. 37 C.F.R. §§ 41.200–41.208.

10. Strictly, many countries employ an "inventive step" standard in addition to the novelty requirement. While not exactly the same, the

requirement that an invention exhibit an inventive step is similar to the U.S. requirement that the invention not be obvious.

11. *KSR International Co. v. Teleflex, Inc.*, 550 U.S., 127 S. Ct. 1727 (2007).

12. Fromson v. Advanced Offset Plate, Inc., 755 F.2d 1549, 1556 n. 3, 225 USPQ 26, 31 n. 3 (Fed. Cir. 1985).

13. See, for example, *In re Rose*, 220 F.2d 459 (CCPA 1955); *In re Reinhart*, 531 F.2d 1048 (CCPA 1976); *Gardner v. TEC Systems, Inc.*, 725 F.2d 1338 (Fed. Cir. 1984).

14. Act of April 10, 1790, 1 Statutes at Large 109.

15. "Fiscal Year 2006: A Record-Breaking Year for the USPTO." Press Release by the U.S. Patent and Trademark Office, December 22, 2006.

16. Robeson, David J., May 2006. Nanotechnology and the USPTO, *The Disclosure*. National Association of Patent Practitioners.

17. USPTO Classification Order 1850.

18. 35 U.S.C. §181.

19. 35 U.S.C. §183.

20. Recently, an alternative to this procedure was introduced that permits applicants to request that the conference be held before filing the appeal brief, with consideration of the issues being based on a much shorter description of the applicant position. This procedure was introduced to account for the fact that in roughly 50 percent of all cases, the other participants in the conference disagreed with the examiner. The objective of permitting the conference to take place before submission of the appeal brief is to avoid preparation of such a detailed document in many circumstances.

21. Spurred by inconsistencies in the way patent cases were being handled, the United States created the Court of Appeals for the Federal Circuit in 1982 to provide a single appellate court that would hear all patent appeals. The objective was to increase consistency and predictability in the way patent law was interpreted, although criticism sometimes still remains that this objective has not been well accomplished. In addition to patent cases, the court hears appeals involving international trade, government contracts, trademarks, certain money claims against the U.S. government, federal personnel, and veteran's benefits.

22. Manual of Patent Examining Procedure, §708.02.

23. Special status for biotechnology applications is available only for "small entities," which are essentially individual inventors, small business concerns having fewer than 500 employees, and nonprofit organizations. 37 C.F.R. §1.27(a). There are certain other advantages available for small entities, the most significant of which is a 50 percent reduction in most fees paid to the patent office.

24. In addition to the technology areas discussed, special status may be afforded to applications in which an applicant is in poor health or is of an advanced age, in which manufacture of the invention is awaiting successful grant of a patent, or where an infringing product is already on the market. Individual requirements must be met for each of these categories and they may also provide a basis for obtaining special treatment for a nanotechnology (or any other) invention.

25. This and a number of other anecdotes describing the early days of the patent system in the United States may be found in Dobyns, Kenneth W. 1994. *The Patent Office Pony: A History of the Early Patent Office.* Fredericksburg, VA: Sergeant Kirkland's Museum and Historical Society.

26. Sandburg, Brenda. July 30, 2001. Battling the Patent Trolls. *The Recorder.*

27. 35 U.S.C. §284.

28. *SCM Corp. v. Xerox Corp.*, 645 F.2d 1195 (2d Cir. 1981); *In re Xerox Corp.*, 86 FTC 364 (1975).

29. Shapiro, C. 2000. Navigating the Patent Thicket: Cross Licenses, Patent Pools, and Standard-Setting. *Innovation Policy and the Economy*, 1:120.

30. Dobyns, Kenneth W. 1994. *The Patent Office Pony: A History of the Early Patent Office*: Fredericksburg, VA: Sergeant Kirkland's Museum and Historical Society.

31. 17 U.S.C. §106.

32. 17 U.S.C. §107.

33. The categories defined in 17 U.S.C. §102(a) are: (1) literary works; (2) musical works, including any accompanying words; (3) dramatic works, including any accompanying music; (4) pantomimes and choreographic works; (5) pictorial, graphic, and sculptural works; (6) motion pictures and other audiovisual works; (7) sound recordings; and (8) architectural works.

34. 17 U.S.C. §102(a)(8).

35. 17 U.S.C. §101.

36. 17 U.S.C. §102(a)(1).

37. 17 U.S.C. §101.

38. 17 U.S.C. §102(a)(2), §102(a)(7).

39. 17 U.S.C. §101.

40. Musical works present a number of difficulties in copyright law, particularly in determining how the general originality requirement is to be evaluated when comparing with other musical works. These issues exist even in conventional applications of copyright law. A more detailed discussion of these issues can be found in *Nimmer on Copyright*, Chapter 2 (Melville B. Nimmer and David Nimmer, Matthew Bender & Company, 2006).

41. The answer to these questions is, in fact, clear. When the U.S. Congress extended copyright protection to architectural works, it defined *building* as something that is designed to be occupied by humans, giving houses, office buildings, churches, and museums as examples. Furthermore, to qualify as a building, a structure must be permanently affixed to land. 57 Fed. Reg. ¶45,307 (October 1, 1992).

42. 17 U.S.C. §905.

43. 17 U.S.C. §906(a).

44. 17 U.S.C. §901(a)(1)(A).

45. 17 U.S.C. §901(a)(1)(B).

46. See, for example, the Canadian Integrated Circuit Topography Act, §2(1), the Australian Circuits Layout Act, §5, or the United Kingdom Design Right (Semiconductor Topographies) Regulations, §2(1). Japan takes an intermediate approach, permitting the use of a "semiconductor material or insulating material" as the base, but still requiring that structures be "designed to perform an electronic circuitry function." Japanese Semiconductor Layout Act, §(2)(1).

47. *Diamond v. Chakrabarty*, 447 U.S. 303 (1980): "A live, human-made micro-organism … constitutes a 'manufacture' or 'composition of matter' within [the Patent Act]."

48. *State Street Bank & Trust Company v. Signature Financial Group, Inc.*, 149 F.3d 1368 (Fed. Cir. 1998), *cert. denied* 525 U.S. 1093 (1999).

49. U.S. Const., Art. II.2,¶1.

50. Strictly, the similarity of composition of the various colas might prove relevant as an evidentiary issue in combination with proof in showing theft of a trade secret. For instance, if there was evidence of access to the composition of Merchandise 7X together with evidence that the composition of some competitor cola was nearly identical to that of Coca-Cola, this combination could be relevant in proving theft of the secret.

51. *Lear Siegler Inc. v. Ark-Ell Springs Inc.*, 569, F.2d 286, 288 (5th Cir. 1978); *Learning Curve Toys Inc. v. PlayWood Toys Inc.*, 2003 U.S. App. LEXIS 16847 (Aug. 18).

52. Publicly traded companies are subject to a variety of different disclosure requirements that prevent them from maintaining certain kinds of financial information as a trade secret. Private companies are generally not subject to such requirements and may permissibly maintain all financial information as a trade secret.

53. I admit that this example is somewhat contrived. First, in expressing the requirement that the secrecy of the information must provide an economic benefit, statutes generously give credit to potential economic benefits in addition to actual economic benefits. The Uniform Trade Secrets Act is illustrative as defining a *trade secret* as something

that "derives independent economic value, actual or potential, from not being generally known to, and not being readily ascertainable by proper means by, other persons who can obtain economic value from its disclosure or use." In the example in the text, the patent covering NanoD's product will eventually expire, at which point the secrecy of the synthesis technique will have actual economic value. It thus seems that the secrecy of the technique always has potential economic value. Indeed, it is difficult to concoct a scenario in which information has some intrinsic economic benefit but its secrecy is valueless. I am unaware of any case to have elucidated the circumstances under which that might be the case, but the distinction nevertheless persists in most trade-secret statutes.

54. U.S. Pat. No. 6,093,302 and U.S. Pat. No. 6,280,595.

55. Nanogen press release dated December 19, 2002.

56. 35 U.S.C. §§ 200–212.

57. 35 U.S.C. §261.

# II

## Regulation

### A.  Delegation of Power to Agencies

At about 4:00 a.m. on March 28, 1979, a failure in the main feedwater pumps of Unit 2 of the Three Mile Island nuclear power station in Pennsylvania precipitated the most notorious accident involving nuclear power in the United States. The failure of the feedwater pumps prevented steam generators from removing excess heat, resulting in a pressure increase in the nuclear portion of the plant. A relief valve opened to release the excess pressure, but failed to close when the pressure normalized, permitting coolant to stream through the valve and allow the nuclear reactor to overheat. This then combined with failures in instruments, which provided incorrect information to operators of the plant. The result was that the operators responded with actions that exacerbated the condition by reducing coolant flow even further. The result was a severe nuclear-core meltdown as the nuclear fuel reached temperatures that ruptured its protective cladding.

Newspaper headlines screamed "race with nuclear disaster,"[1] and "nuke leak goes out of control,"[2] while newspaper photo captions of the area wondered "Three Mile Island ... Beauty or Holocaust?"[3] Much of the reaction was overly sensationalist, but the accident had a profound effect on the regulation of nuclear power generation in the United States in the decades that followed. This effect was exacerbated by the accident at Chernobyl in the USSR on April 26, 1986, when a violent explosion at Unit 4 of that nuclear power station destroyed its reactor. That accident produced a plume of radioactive dust that settled over much of Europe and Asia and whose effects were detected in the Americas. To this day, researchers continue to uncover biological and environmental anomalies attributed to radioactive fallout from that accident.

There have been no accidents involving nanotechnology that have had as vivid an impact on public perceptions as these nuclear accidents have had. The public demand for regulation of nanotechnology has accordingly not been nearly as fervent as the demands that were made for regulation of nuclear industries. But prompted by concerns about potential biological impacts on human beings and about environmental impacts, there have still been calls from many quarters for limits to be circumscribed on the use of nanotechnology in a more systematic way. This part of the book is not intended to discuss what those limits should be; such policy considerations are more effectively analyzed in the companion books in the Perspectives in Nanotechnology series about the environment by Jo Anne Statkin and about ethical considerations by Deb Bennett-Woods. Instead, this part focuses on the legal mechanisms by which regulations are implemented and on those regulations that are already in force to limit the use of nanotechnology.

## 1. Regulatory Agencies

Government regulation is ubiquitous. There is hardly an area of anyone's personal or professional life that is not impacted in some way by numerous forms of regulation. Regulation is manifested in a variety of economic forms as price and wage controls, through tax and tariff policies, as well as through the imposition of standards designed to promote health, safety, environmental, and other objectives. Frequently, these different aspects of regulation are intertwined with licensing provisions so that only those who have been granted a government sanction are permitted to perform some function, usually upon a demonstration of compliance with standards.

While it is theoretically possible for governments to regulate directly by passing legislation, this would be a wholly impractical approach. There are simply too many activities that society has deemed worthy of being controlled for any reasonably sized legislature to be adequately informed and to design adequate specifics that the regulations take. Instead, authority is delegated by legislatures to agencies that are established to develop an expertise in specific areas. This authority is then used to promulgate rules in a quasi-legislative fashion that establishes a framework governing that area. For instance, regulation of nuclear industries in the United States has been relatively limited to development and implementation by the Atomic Energy Commission (AEC) from 1946 to 1974 and by the Nuclear Regulatory Commission (NRC) since 1975. Both of these commissions were created by legislative acts, with the AEC being created by the Atomic Energy Act and the NRC being created by the Energy Reorganization Act, which also assigned responsibility for some activities of the AEC to the Energy Research and Development Administration.

But most industries are subject to regulations that arise from a variety of different agencies. In the United States, there are some one hundred agencies that have been granted the power to impose regulations within certain defined areas. Some of the agencies that might have an impact on the manner in which nanotechnology is developed or marketed include the National Institutes of Health, the Forest Service, the Farm Service Agency, the Agency for Toxic Substances and Disease Registry, the Food and Drug Administration, the Health Resources and Services Administration, the Consumer Product Safety Commission, the Environmental Protection Agency, the Occupational Safety and Health Administration, and the United States Office of Government Ethics. The regulatory framework within which nanotechnology researchers and companies will—and already do—operate is complex. At the moment, there is no agency devoted exclusively to issues surrounding the use of nanotechnology, and it is an interesting exercise to consider what might be involved with such an agency.

In many respects, agencies act with the same kinds of powers that are claimed by sovereign governments. They have a clear legislative function embodied in their rule-making power to promulgate regulations that dictate standards of quality for products or of behavior for individuals and organizations. They also have an executive function that permits them to investigate and prosecute violations of their regulations. And they have a judicial function that allows them to make adjudications that those rules have been violated. At least with the organization of government in the United States— which is structured fundamentally to separate these powers among different branches of government—questions exist about the propriety of such a structure. The very purpose of separating these powers is to limit the risk of corruption that exists when power is concentrated.[4] Indeed, even in countries with a less formalized separation-of-powers structure, the fundamental issue of how to constrain regulatory bodies from excess exercise of their powers is important.

It is generally accepted that such delegations of authority are permissible, provided that certain constraints are met. The creation of an agency with regulatory power should be made with some intelligible standard and a set of substantive and procedural controls that limit the power of the agency.[5] Of particular relevance is the availability of judicial review of agency actions by a court; this permits the actions of the agency to be evaluated against the legislative intent in establishing the agency.

## 2. Rulemaking

The U.S. Administrative Procedure Act, which defines a process by which agencies may promulgate regulations and establishes a procedure to provide court access for reviewing agency decisions. When an agency wishes to make any kind of change to its body of rules—whether that be to promulgate a new rule, abolish an old rule, or make an amendment to an existing rule—it must follow the procedure outlined in the act. A principal objective of this procedure is to ensure that there is some opportunity for the public to provide input relevant to the proposed change.

This procedure can take one of two forms, referred to as *"informal"* and *"formal"* rulemaking. The informal process is also sometimes referred to as a *"notice and comment"* process, identifying its two basic components.[6] With this process, the agency publishes its proposed rule changes and establishes a time period over which members of the public are permitted to offer comments. This time period is usually thirty, sixty, or ninety days, and comments are typically received from a wide range of sources. It is not uncommon for individual members of the public to provide comments that express their views, although the more substantive comments tend to come from organizations that have an interest in the particular rules. For instance, when the Environmental Protection Agency (EPA) proposes a rule change that affects nanotechnology, professional scientific and engineering societies involved in nanotechnology research might generate and submit comments, as might nanotechnology corporations that would be affected by the rule changes. Legal societies and associations are frequent sources of comments, and they typically offer a perspective on how the rule changes will successfully or unsuccessfully integrate with other regulations.

After the comment period has closed, the agency reviews all the comments and issues final rules, usually providing a detailed response to the comments that were received and explaining how the proposed rules were modified to address the comments. This basically concludes the informal process, with the final rules then forming the current set of regulations. The ability of the public to provide input in the informal process is limited to submitting comments on proposed rules. In some cases, members of the public may still object to the form of the final rules. While it is possible to challenge such final rules as an overreaching of the authority of the agency, they will stand as enforceable regulations if they are within the proper scope of the agency's power.

In actual practice, this comment procedure is very valuable. It frequently highlights ambiguities in the form of the rules as proposed and often highlights concerns that the agency failed to appreciate fully when formulating its proposal. Currently, the ability of the public to provide comments is greater than it has ever been. All rulemaking proposals for all U.S. agencies are published in the Federal Register. Organizations that are governed by specific regulatory agencies thus routinely check the Federal Register to identify any proposed rules that might affect them. The Internet also provides access to all proposed rules for U.S. agencies at www.regulations.gov. Many agencies additionally publish received comments on their individual Web sites. The World Wide Web provides a wonderful mechanism by which interested members of the public can read the comments that have been submitted by others, permitting them to elaborate on issues that might not have been ideally expressed or to offer a different perspective on the same issue.

Much less widely used is the formal rulemaking process.[7] This process is essentially used only when the agency is precluded from using the informal process because of some specific congressional directive, usually embodied within some other statute. The formal process requires that a trial-like hearing be held, overseen by an administrative law judge. During the hearing, interested persons are to have an opportunity to testify and be cross-examined before a rule can be promulgated. A number of notorious proceedings have resulted in the formal rulemaking process having been largely discredited. Most famous are the peanut butter proceedings that were held to determine whether a product labeled *peanut butter* must have at least 87 percent or 90 percent peanuts. The formal process took nine years and resulted in a transcript having more than 7500 pages.[8]

The regulatory authority that is delegated to agencies is thus significant. These agencies have the power to promulgate rules that can have profound effects on industries. The major constraint on agency activities is the ability of the legislature to reassume some of the power it has delegated. Through legislation, there are a number of ways in which the power of agencies can be limited: specific agency decisions can be overruled by the targeted legislation; the scope of authority of an agency can be altered by more general forms of legislation; it even remains an option to dissolve the agency if circumstances warrant.

## 3. Use of Information by Agencies

Even with their regulatory powers, many agencies have more recently begun to engage in a practice that is sometimes derisively referred to as *"regulation by information."* Rather than regulate certain aspects of an industry directly—and thereby have to adhere to the requirements of the Administrative Procedure Act—some agencies simply publish information in a manner believed to be influential on industries. A commonly cited example of this practice is promulgation of the Toxics Release Inventory by the EPA. Industry and federal facilities are required to provide information to the EPA detailing releases of specific toxic substances and how they have been managed as wastes. This information is then maintained and published by the agency. While it had historically been published only in aggregate form, a more recent trend has been to publish it with the information tied directly to the reporting entities.

The rapid increases in the ability to disseminate information, particularly with the development of the Internet, have increased the impact of the availability of such data. The concern surrounding the Toxics Release Inventory among some industries is that its publication with the imprimatur of a federal agency is designed to embarrass them into taking actions that the agency has not required. In many instances, industries complain that exertion of this pressure seems improper because it is being applied without the agency having complied with the necessary procedures to impose regulation.

Of perhaps even more relevance to nanotechnology companies are technical determinations that are disseminated by agencies. The EPA provides another example in this area in the form of its Integrated Risk Information System. This is a database that describes the various effects that exposure to environmental substances can have on human health. The data reflect a consensus view of EPA scientists, but there are sometimes charges that the information is inaccurate or is based on scientifically defective studies.

## 4. Information Quality

Recently, the U.S. Congress used its legislative power to respond to these concerns surrounding the use of information as a surrogate for direct regulation. The Data Quality Act (sometimes referred to as the "Information

Quality Act") has been characterized as a "nemesis" of regulation.[9] Passed as a two-sentence rider to a spending bill in 2000,[10] the Data Quality Act has a surprising potential for significant impact on many of the activities of regulatory agencies. It essentially required that agencies develop procedures to review and substantiate the quality of data before dissemination. It also required that a mechanism be established to challenge information that is disseminated so that it can be withdrawn or corrected.

In many respects, these requirements sound reasonable. But they have been criticized as a subversion prompted by industry lobbyists to insert onerous and burdensome requirements on the use of information by agencies as a means of hampering their effectiveness. This allegation is made especially in the context of challenges to scientific and technical conclusions publicized by regulatory agencies. Guidelines developed pursuant to the Data Quality Act require that stricter standards be applied to information that is considered "influential," in the sense that dissemination of the information can reasonably be expected to impact public policy evaluations. Much scientific and technical information is "influential" in this way.

For such influential information to be disseminated by a regulatory body, it must meet strict reproducibility standards. Influential information that is related to human health or safety, or to the environment, must be based on "the best available, peer-reviewed science and supporting studies conducted in accordance with sound and objective scientific practices."[11] Again, these requirements sound eminently reasonable. But those who criticize the Data Quality Act charge that these standards require a much greater level of certainty than is conventional in scientific endeavors before action is taken. After all, the very nature of science embraces the concept that no piece of information is sacrosanct; any doctrine or conclusion is always subject to reevaluation in light of new data. And meeting the strict reproducibility standards imposed on information used in developing regulations may strain the financial capacity of the regulatory body so that it is effectively rendered unable to act.[12] In essence, this counterargument asserts that the application of the Data Quality Act confuses data that are incomplete with data that are of poor quality, and that it may still frequently be appropriate for regulatory action to be taken even on the basis of data that are not fully complete.

It is true that many uses of the Data Quality Act since its passage have been clearly designed to promote industry interests. The Nickel Development was among several parties that used the act to challenge a government report on nickel hazards. The Salt Institute was a party that challenged data relied on by the National Institutes of Health in developing human salt-intake recommendations. The Sugar Association was among several sugar interests that challenged the Agriculture Department and Food and Drug Administration regarding recommended limits on dietary sugar consumption by humans. And so on.

At the same time, the Data Quality Act has also been used by a number of public interest groups, although such uses have been less frequent than uses by industrial interests. Public Employees for Environmental Responsibility used the act to challenge data used by the Fish and Wildlife Service in developing a recovery plan for the Florida panther that they found inadequate. The sexual-health groups Advocates for Youth and Sexuality Information and Education Council of the United States used the act to challenge data used by the Administration of Children and Families of the Department of Health and Human Services in promoting abstinence-only sexual programs. The Department of Health and Human Services has also been faced with a challenge brought by Americans for Safe Access as promulgating inaccurate information regarding medical uses of marijuana to support its classification of marijuana as a Schedule I drug.

Irrespective of the motivation behind passage of the Data Quality Act, it remains a provision that has great potential to affect the way nanotechnology is regulated. Nanotechnology is a discipline that will see development of a great deal of scientific and technical information that will be subject to the act. In addition to being able to provide comments in a relatively passive fashion to respond to proposed regulatory changes under the notice-and-comment provisions of the Administrative Procedure Act, the Data Quality Act now gives nanotechnology interests the ability to bring more direct challenges. At the moment, it is possible to imagine challenges being brought under the Data Quality Act both by those who are generally supportive of the use of nanotechnology and by those who fear some of its potential consequences. As a technology that is still in a somewhat nascent state, increasing numbers of scientific studies on the health and environmental effects of nanotechnology are likely in the next several years. The results of long-term studies are among those that are particularly deficient at the moment but which will undoubtedly be delivered in the years to come. And the pace of reporting results is likely to increase.

Whenever these results are disseminated by a regulatory agency, they may be actively challenged. For instance, studies that show harmful effects to human health or to the environment may be challenged by those with commercial interests in promoting nanotechnology. Corporations that produce products using the specific technology being discussed are likely to use the act to challenge the adequacy of the study. But studies that show certain nanotechnologies to be benign are also likely to result in challenges when they are disseminated by agencies to support their decision not to regulate. There are already a number of advocacy groups that believe that the current level of uncertainty about the health and environmental effects of nanotechnology warrants a cautious approach. While commercial interests argue that nanotechnology use should be largely unregulated until some demonstrable harms are identified, these groups argue that it should be strictly regulated until its safety has been unquestionably established. Early studies showing no adverse effects to nanotechnology are susceptible to challenges that their protocols were insufficiently comprehensive to provide a conclusive demonstration of safety.

There are four ostensibly different bases on which information can be challenged, although the Data Quality Act itself does not provide any precise definition of how they differ: quality, utility, objectivity, and integrity. Some effort has been made to clarify what these terms mean, and therefore how a challenge under the act may be structured, in guidelines prepared by the U.S. Office of Management and Budget.[13] The term *"quality"* is intended to be overarching and to encompass the other terms as constituents. The most controversial constituent is the requirement of *"objectivity,"* which is explained as "focus[ing] on whether the disseminated information is being presented in an accurate, clear, complete, and unbiased manner, and as a matter of substance, is accurate, reliable and unbiased." This is the basis on which most challenges are organized, which assert that the information is somehow inaccurate, incomplete, or reflects some bias. The *"integrity"* constituent is directed at whether the information is sufficiently protected from falsification or corruption and the *"utility"* constituent merely refers to whether the information is useful.

Even with all of this capability to use the Data Quality Act, there remains an open question about the possibility of judicial review. After a party initiates a challenge to information under the act, it is still the agency that decides how to respond. While the Data Quality Act requires that there be an administrative mechanism for parties to seek correction or withdrawal of disseminated information, it is still the agency that implements this procedure. What is unclear is whether an appeal can be made to a court if the agency persists in disseminating information that is believed not to comply with the Data Quality Act. Traditionally, courts have not reviewed the quality of information in this way because it was not dispositive in any way of the legal rights or obligations of parties. But passage of the Data Quality Act may have changed this perspective, raising the question of whether nontechnical judges will now be put in a position of having to assess the "quality"—in all its various dimensions—of information that can be highly technical.

The various considerations outlined earlier—the delegation of power to regulatory agencies, the methods by which they establish and promulgate rules, and the manner in which they disseminate data—are all generally applicable to regulation in all areas in the United States. But it is fitting to single out nanotechnology for special comment. As a technology that embraces a diverse array of scientific and engineering disciplines, it is likely to be subject to a variety of different ways in which these considerations apply. Only some of the regulatory agencies that will be involved with nanotechnology were mentioned earlier, and each of them may implement procedures in somewhat different fashion to comply with the mandates of the Administrative Procedure Act and Data Quality Act. Furthermore, nanotechnology is more likely than more mature technologies to raise new issues that have not been as fully considered in the past. These new issues will require innovative regulatory approaches and will engender a variety of unique challenges to those approaches.

## Discussion

1. The text discusses "formal" and "informal" processes for rulemaking in the United States. However, it is worth recognizing that there are a number of exceptions to the rulemaking process in which agencies are able to promulgate new rules with no public input whatsoever. These excepted cases include rules related to military or foreign affairs; rules related to certain proprietary matters such as loans, grants, and benefits; and rules where obtaining public input is impracticable or contrary to the public interest.[14] This last provision has been found to apply when the purpose of a rule would be frustrated by the delay needed to obtain public input or where the rule is so administratively trivial that public input would be valueless.

   Construct several hypothetical examples of rules involving nanotechnology that would be exempt from public participation. Do any of your examples strike you as inconsistent with the fundamental structure of a participatory government? What changes, if any, would you make to the exemption requirements so that they are more consistent with your personal philosophy of representative government? Do you think any specific changes targeted at nanotechnology are desirable or warranted? Or can issues you identify with your examples be accommodated with more general changes to the structure?

2. Investigate the results of at least three challenges to information dissemination by regulatory agencies under the Data Quality Act. Try to find challenges initiated by both industry groups and public-interest groups to include in your sample. Do these results support the view that the act is interfering with legitimate regulatory activity? How difficult is it for you to separate your own political bias regarding the issues underlying the challenges from how you judge these results?

3. Nanotechnology researchers who are funded by an agency may find themselves subject to the Data Quality Act when the funding agency either requires publication of the results as part of the funding grant or has the authority to review and approve it before publication. Such dissemination is referred to as *"sponsored distribution"* and triggers the quality standards of the act. How might this affect the manner in which such researchers conduct their research? How might it affect the nature of funding grants?

4. In a study reported in March 2004, researchers at Southern Methodist University in Dallas, Texas, exposed nine juvenile largemouth bass to water-soluble buckyballs $C_{60}$ at a dose of 500 ppb. Within forty-eight hours, a seventeenfold increase in cellular damage was observed in the brains of the fish in the form of lipid peroxidation when compared with nine unexposed fish.[15] Suppose this study is

disseminated by the Environmental Protection Agency with other evidence to support a proposal to restrict the release of fullerenes into the environment. Investigate research subsequent to this study and use it to formulate the outlines of a challenge to the proposal under the Data Quality Act.

## B.   Examples of Regulation of Nanotechnology

### 1.   Health and Safety Regulations

On June 1, 1992, the *New York Times* published an opinion piece by Sheldon Krimsky titled "Tomatoes May Be Dangerous to Your Health," in which Dr. Krimsky criticized the exemption of genetically engineered crops from certain levels of review by the Food and Drug Administration. This article was met with a response by Paul Lewis, in which the term *"Frankenfood"* was coined to refer to food derived from such crops:[16]

> "Tomatoes May be Dangerous to Your Health" (Op-Ed, June 1) by Sheldon Krimsky is right to question the decision of the Food and Drug Administration to exempt genetically engineered crops from case-by-case review. Ever since Mary Shelley's baron rolled his improved human out of the lab, scientists have been bringing just such good things to life. If they want to sell us Frankenfood, perhaps it's time to gather the villagers, light some torches and head to the castle.[17]

This rather derisive characterization of food derived from genetically modified organisms quickly became popular, reflecting a fairly widespread distrust of the safety of such food. Numerous, highly visible protests against government decisions permitting the sale of such food followed, being notably strong in Europe. The result was a long confrontation between scientists and those who were concerned about the potential harms that could result from ingestion of food derived from genetically modified organisms.

But in fact, human beings have been engaged in genetic manipulation of organisms for millennia. Most of this genetic manipulation has taken the form of domestication of plants and animals—and the evidence of genetic manipulation arising from domestication was one of the important pieces of evidence put forth by Charles Darwin in his thesis in *The Origin of Species*. Examples of such genetic manipulation by human beings abound. Perhaps the most commonly advanced example is dogs, which share a common ancestor with modern wolves. Dogs have been bred to have specialized domestic roles, with those breeds that become pets having been bred to have youthful physical characteristics—a short muzzle, small teeth, round eyes—and to have a docile demeanor. Those bred to act as herd or hunting dogs emphasize other characteristics—notably the instinct to stalk prey—but remain much more docile than their wolf cousins.

This kind of domestication illustrates a certain acceptance by human beings of their ability to engage in genetic manipulation. While much of that kind of domestication was done without a good understanding of the underlying genetics, more modern examples evidence a more pragmatic acceptance of our ability to control the characteristics of species genetically. In Russia in the 1950s, for example, silver foxes that were farmed for fur were savage animals. They managed poorly in farm captivity, often dying from anxiety. But with a deliberate program of selective breeding, their demeanor was changed in only twenty years to a group of considerably tamer animals that also had more desirable features in their coats.

Other examples more directly related to the "Frankenfood" issue are to be found in food production. In these cases, genetic diversity is often sacrificed in favor of highly valuable strains of species for food production. For example, it is startling to learn that the entire soybean crop in the United States, which amounts to some 60 million tons, is descended from a mere dozen soybean strains that were collected in northeastern China. The same predilection for reproducing the genes of superior specimens is evident in beef and dairy production. When the Dutch Holstein Friesian bull Sunny Boy died in 1997, he had sired some 2 million calves, resulting in a significant and deliberate impact by human beings on the genetic makeup of the world's cattle population.[18]

It is within this context that fears about "Frankenfood" should be judged. In many ways, it is the increased level of control over genes that gives rise to fears about the potential dangers of genetically modified organisms. But when viewed against the backdrop of how intrusively human beings have always influenced the genetic makeup of other species, the concerns about genetic modification seem somewhat irrational. To be sure, there may be certain risks associated with genetic manipulation directly at the level of DNA, but the notion that genetic modification of species to suit human purposes somehow represents an abomination of nature is surely overblown. And yet, the challenges faced in gaining public acceptance of using genetically modified organisms as food sources acts as a sharp reminder of the potential challenges to be faced in gaining public acceptance of foods produced using nanotechnology.

### i. The U.S. Food and Drug Administration

Like so many other areas, the ability to apply nanotechnology to food products is in many respects limited only by the imagination of researchers. The potential for application of nanotechnology to food production can occur in virtually any stage, including during the original agriculture of food, the processing of food, the packaging of food, and even incorporated as a form of food supplement. At any stage of this production chain, there is the potential for nanoscale materials to become integrated with the food itself and ultimately ingested by human beings. For instance, nanoscale detectors may be used to monitor enzyme interactions in farmland, nanoscale capsules may be used in the delivery of pesticides or for delivering growth hormones and vaccines, nanosensors may be used for monitoring soil conditions or for detecting pathogens, and other nanoscale devices may be used in other agricultural applications. As part of widescale food processing, nanoscale devices may find application in the form of nanocapsules to improve the availability of certain neutraceuticals in food products or may encapsulate flavor enhancers, nanotubes may be incorporated into food as gelation agents, and nanoemulsions may be used to improve the dispersion of nutrients. The packaging of food may benefit from the use of biodegradable nanosensors that test for temperature and moisture content, from antimicrobial coatings formed from nanoparticles, and from films that incorporate nanostructures to render them lighter, stronger, and more heat resistant. Examples of ways in which nanotechnology may be used with supplements include the preparation of supplements as nanosized powders to improve nutrient absorption and the use of coiled nanoparticles to deliver nutrients to cells in a more efficient manner without impacting the color or taste of food.

These are just some of the many potential uses of nanotechnology related to food. In the United States, the regulation of food is handled by the Food and Drug Administration (FDA), which is also responsible for the regulation of several other kinds of products, including drugs, certain medical devices, various biologics such as vaccines and blood products, animal feed and drugs, cosmetics, and some products that emit radiation such as cellular telephones, lasers, and the like. While the introductory description earlier ocuses primarily on food products, it is illustrative of another approach to regulation to consider all of the products regulated by the FDA collectively.

The origins of the FDA can be traced to a meeting of eleven physicians in 1820 to establish a compendium of standard drugs in the United States known as the U.S. Pharmacopeia. Since that time, it has grown into a large organization that regulates more than $1 billion of products that account for some 25 percent of all consumer dollars spent by Americans. The number of employees is currently approaching 10,000 located in more than 150 offices throughout the country, of which some 2100 are scientists who work in any of about 40 laboratories. More than 1000 inspectors employed by the FDA visit more than 15,000 facilities a year to ensure compliance with regulations enforced by the agency. These inspections involve the collection of about 80,000 product samples for examination. In a typical year, about 3000 products are found by the FDA to be unfit for consumers and are consequently removed from the market. This is in addition to some 30,000 import shipments that are detained annually at an entry port because of apparent violations by the products.

The regulatory authority of the FDA is based on the Federal Food, Drug, and Cosmetic Act (FDCA), as well as on a number of public-health laws. Which of these particular laws applies to any given product is dictated in large measure by the way in which the product is classified. Classification of nanotechnology products under this scheme has the potential to present certain challenges. And a basis for disputing the application of certain authority with respect to a particular nanotechnology product may well be that the classification is incorrect. When successful, such a strategy may result in the way in which a particular nanotechnology product is regulated becoming more or less stringent, with some public-interest challengers perhaps being more likely to favor more stringent regulation and corporate interests in developing and selling the products perhaps favoring more liberal regulation.

The FDA is organized into six "centers" that are responsible for products that fall within particular classifications, as well as two "offices" that perform more administrative functions (Figure II.1). The centers are the Center for Biologics Evaluation and Research (CBER), the Center for Devices and Radiological Health (CDRH), the Center for Drug Evaluation and Research (CDER), the Center for Food Safety and Applied Nutrition (CFSAN), the Center for Veterinary Medicine (CVM), and the National Center for Toxicological Research (NCTR). The two offices are the Office of the Commissioner, which

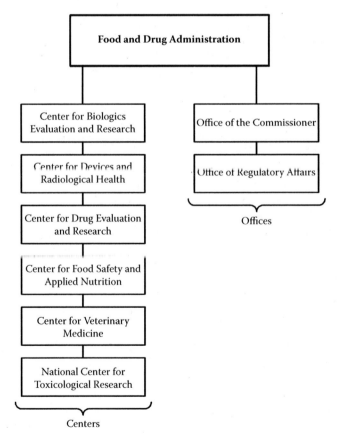

**FIGURE II.1**
Organization of the Food and Drug Administration.

provides management functions for the agency, and the Office of Regulatory Affairs (ORA), under whose auspices operate the scientists who perform research, the consumer safety officers who conduct inspections, and the public affairs specialists who provide a more public face for the agency. Roughly one third of all FDA employees operate under the authority of the ORA.

It is worth reviewing the general scope of authority for each of the centers to better understand how conflicts may arise when various nanotechnology products are considered.

### ii.   The Center for Biologics Evaluation and Research

The CBER is responsible for the regulation of *biological products*, which is an umbrella term that includes many forms of products used in the treatment of human beings. The term is not without some ambiguity, but it is frequently used to distinguish from chemical drugs that are synthesized chemically. Examples of such products include allergenics used by physicians to identify allergies in humans; blood and blood components used in transfusions; gene-therapy products used to replace genetic material in human beings; human tissues used in transplants, such as skin, ligaments, and cartilage; xenotransplantation products in which animal tissue or organs are transplanted into human beings; and vaccines. In addition to the regulation of such biologics, the CBER is responsible for the regulation of certain related medical devices that are used to safeguard biologics.

Any of these has the potential to make use of nanotechnology—nanoscale materials may conceivably be incorporated into blood products, used in gene-therapy techniques, incorporated into human or animal tissue, and so on. But such uses are, at least at the moment, somewhat speculative. Where nanotechnology has clear application under the authority of CBER is in the regulation of medical devices with the appropriate functionality. Such devices may be regulated pursuant to authority under the FDCA, with biologics themselves being regulated according to both the FDCA and the Public Health Service Act. Authority for regulation under both acts is largely a function of the fact that the definition of *"drugs"* under the FDCA is expansive enough to include most biologics.

### iii.   The Center for Devices and Radiological Health

The CDRH is in some respects more directly involved with the regulation of devices. The scope of what is encompassed by the definition of *"medical device"* under these auspices is very broad. "Devices" can range from something as simple as a tongue depressor to a very complex piece of robotic machinery that performs remote surgery and involves the coordination of numerous complex functions that are controlled by sophisticated software. Nanotechnology is a ripe discipline for incorporation into many medical devices. The ability to use MEMS and NEMS structures in medical devices may contribute to further miniaturization of devices, rendering them considerably less invasive than many currently used devices.

The CDRH itself classifies devices into one of three categorizations that reflect a judgment on the level of risk presented by a particular device. Class I devices are those that present the lowest level of risk and are subject only to "general controls." Class II devices present a greater level of risk and are subject to "specific controls." Those devices that present the greatest level of risk—Class III devices—are subject to relatively stringent evaluation, requiring adherence to various strict conditions on manufacturing of the devices and requiring explicit and informative labeling.

But in addition to regulating medical devices, the CDRH also regulates devices that emit radiation. Some devices will fall within both categorizations, but most devices that emit radiation will not have particular medical applications. And the term *"radiation"* itself is relatively broad, encompassing not only electromagnetic radiation but also acoustic radiation. Examples of radiation-emitting devices that are regulated by the FDA through the CDRH thus include not only such devices as tanning lamps, welding equipment, and remote-control devices that emit electromagnetic radiation, but also sound amplifiers, sonic drills, and sonic-based jewelry cleaners, among thousands of other products.

### iv. The Center for Drug Evaluation and Research

The CDER is responsible for the evaluation of drugs, both those that are available only by prescription and those that are available over the counter. In evaluating drugs, the FDA uses a generally accepted sequence of tests. First, the effects of chemicals that may make up a drug are tested in laboratory settings on cell cultures or other compositions that might be involved in a particular disease. If these tests prove promising, then the chemicals are tested in at least two different nonhuman species that act as models to assess the action of the chemical in a biological system. It is only after these initial steps that a potential new drug is tested in humans according to a three-phase protocol. In Phase I, tests are made for adverse effects on a small group of healthy volunteers. If there are no adverse effects—or the effects are acceptably minor—the effectiveness of the drug on the relevant disease or condition is tested in patients suffering from that disease or condition in Phase II. If the drug shows adequate effectiveness, it proceeds to Phase III, which attempts to define specific criteria for use of the drug—that the drug is safe and effective when used in appropriate doses. The Phase III studies are intended to be comprehensive, involving potentially many thousands of patients who participate in a study that permits comparison with a control substance.

This basic structure is expected to be followed with drugs that are developed using nanotechnology. Further discussion of this is provided in a later section, but the effect of nanostructures will first be evaluated in laboratory settings on cultures, then in animal models, and then using the three-phase protocol for human studies.

### v.   The Center for Food Safety and Applied Nutrition

The regulation of food products by the CFSAN is one of the largest roles played by the FDA. Approximately 80 percent of all food consumed in the United States falls under the regulatory authority of the CFSAN, the exceptions being meat, poultry, and egg products, which are instead regulated by the U.S. Department of Agriculture. As in other areas, the potential uses of nanotechnology in food products is in certain respects only limited by the imagination of researchers working in the field. What is especially troubling to much of the public is that proposals for the use of nanotechnology with food products are not limited to incidental items such as packaging or to processing methodologies, but are proposed to be included as part of the food to be ingested. For instance, they may be incorporated as food colorings, preservatives, and other additives that arguably improve the food to be consumed by making it more palatable, more resistant to spoilage or infection, and so on. And even when used on incidental products, like when anti-microbial nanotech films are coated on packaging, there remains a certain likelihood of ingestion.

Growth in the incorporation of nanotechnology in the food industry is accelerating. At the moment, only some $500 million (out of a total $3 trillion global food market) are directed to nanotechnology, but that number is expected to grow by more than a factor of ten within the next five years. The authority of the CFSAN to regulate the use of nanotechnology in the food industry is derived from generally more statutes than is the case for the other centers of the FDA, reflecting the high level of importance placed on ensuring the safety of the food supply in the United States. Some of the statutes on which the CFSAN may draw for regulatory authority of course include the FDCA, but also include the Federal Food and Drugs Act, the Federal Import Milk Act, the Public Health Service Act, the Infant Formula Act, and the Dietary Supplement Health and Education Act, in addition to the Fair Packaging and Labeling Act and the Nutrition Labeling and Education Act. An interesting question (see the first item in the next Discussion section) relates to whether it is appropriate to require labels identifying foods that contain nanoparticles.

### vi.   The Center for Veterinary Medicine

The various centers of the FDA that have been discussed thus far have been concerned with regulations designed to protect the safety and welfare of human beings. But many of the same issues apply equally to animals, which fill a variety of domestic and farming purposes. Such animals consume food that has been supplied through sophisticated food chains that largely mimic the food chains used to supply food for human beings. And they ingest drugs that have been supplied by veterinarians, not only in the treatment of disease but also as an adjunct to certain agricultural practices. The CVM is responsible for the regulation of food and drugs as developed specifically

for animals. The existence of a separate center for such purposes is not especially relevant to a discussion of the regulation of nanotechnology, except to
keep in mind that regulation need not be concerned only with the welfare
of human beings; there are other species that are also potentially affected by
nanotechnology products that regulation seeks to protect.

### vii.   National Center for Toxicological Research

The NCTR is somewhat different from the other FDA centers. This center is
charged with providing a source of scientific expertise on which the FDA can
draw in making regulatory decisions. As such, the center is more involved
in undertaking scientific research, particularly on understanding the manner in which toxicity of certain substances is expressed in biological organisms; it is not directly involved in deciding how to apply the results of that
research within the spheres of responsibility of the other centers.

### viii.   Classification of Medical Products

The issue of how to classify various nanotechnology products arises in
deciding which of the different centers has jurisdiction over the product. The
determination can be critically important in defining the scope of regulation
that a particular product is subject to because of the different approaches
taken by the different centers. Of course, the classification system used by the
FDA reflects the division of the centers as described above. Without forget-
ting that the FDA regulates other products also, a discussion of the classification issues may usefully begin by considering the different classifications of
medical products: drugs, devices, biologics, and combination products. The
first three of these have been discussed, but "combination products" perhaps
deserve some additional comment. Basically, a *combination product* is exactly
what the term suggests—it is some product that has multiple components
that fall within more than one of the other classifications: combination products are drug + device, drug + biologic, device + biologic, or drug + device +
biologic.

Essentially, classification of a product as a "drug" means that it will be
regulated by the CDER; classification as a "device" means it will be regulated
by the CDRH, and classification as a "biologic" means it will be regulated by
the CBER. When a product is a combination product, it is the "primary
mode of action" of the product that governs. This means that the manner in
which a combination product is regulated may very well hinge on what is
determined to be its "primary mode of action." A drug-device combination
product could potentially be regulated according to the requirements of the
CDER or the CDRH.

The classification of combination products is a relatively recent development at the FDA, perhaps reflecting the fact that the original classifications
of drug, device, and biologic were by themselves insufficiently flexible to
accommodate various technical advances that blurred these distinctions.

The classification of "combination product" was created by Congress in 1990 and separate (two-way) agreements between the different centers were established to simplify determination of the primary mode of action and assignment to the correct center. But even so, there have been a number of controversies over jurisdictional assignments of combination products, not only initiated by manufacturers of the products but also initiated by one or more of the centers themselves. In 2003, an Office of Combination Products was established to make the assignments and to resolve disputes. It is generally expected that this office will improve the effectiveness of the assignments of combination products to specific centers.

What is interesting about nanotechnology is that it very naturally blurs some of the bases for distinguishing between the different classifications, which were developed from macroscopic properties and behaviors. When considering the manipulation of matter at the microscopic level, what distinguishes a "drug" from a "biologic"? And in what way are the intrinsic chemical interactions between matter at that level distinct from its manipulation with a "device"? These are questions that currently have no clear answer, and yet the specific regulations that will govern the use of nanotechnology very much depend on how specific products are to be classified.

The FDA has recognized the challenges involved with regulating nanotechnology and initiated a number of concrete steps to address them. In particular, the FDA actively participates in a number of committees and working groups, including the Nanoscale Science and Engineering Technology (NSET) Subcommittee of the National Science and Technology Council Committee on Technology. It is a cochair of the NSET Working Group on Nanomaterials Environmental and Health Implications, which works to define methods for testing and evaluating the safety of nanotech products. It also participates in programs run by the National Institute of Environmental Health Sciences and the National Toxicology Program to evaluate the toxicity of nanomaterials.[19]

In addition to participation with these external groups, the FDA has a number of internal efforts directed at ensuring that it is well informed about nanotechnology issues and in a position to promulgate regulations that are meaningful. Perhaps foremost of these internal efforts is the formation of the Nanotechnology Task Force in August 2006. This task force is "charged with determining regulatory approaches that encourage the continued development of innovative, safe and effective FDA-regulated products that use nanotechnology materials."[20] The mandate of the task force is relatively broad, requiring that it hold a public meeting and engage in a process that assesses the current state of scientific understanding of nanotechnology and both to evaluate the effectiveness of existing regulatory approaches and to explore how nanotechnology might be used in developing safer products.

In addition to this task force, the individual centers and the Office of the Commissioner engage in efforts to familiarize those employed by the agency with issues related to nanotechnology. For instance, the CDER has a

Nanotechnology Working Group that seeks to develop a better understanding of how nanotechnology impacts regulatory issues within its group. In addition, the Office of Science and Health Coordination, which is a part of the Office of the Commissioner, helps to coordinate a variety of discussions within the agency, some of which relate to nanotechnology.

### Discussion

1. One of the issues raised in the main text involves labeling of food products that may contain nanoscale structures. What arguments can you think of in favor of such labeling? What arguments can you think of against labeling? Research the controversy over labeling of foods derived from genetically modified organisms. What can you say about the level of consensus reached regarding labeling? Do you think that considerations similar to those that applied in that controversy also apply to nanotechnology? Are there any significant differences between the technologies that argue in favor of a different result?

2. The main text suggests that regulation of veterinary products is largely similar to the regulation of products intended for human use. Based on your general experience, do you agree that this is the case? In particular, is it possible that regulatory authorities might tolerate greater intrusion of impurities in food chains to animals than to humans, or might tolerate more unknowns about the effects of drugs on animals than on humans? If so, do you have any concerns about the different treatment of nanotechnology as it applies to human consumption and animal consumption? Why do you have such concerns and what would you propose to address them? Is your proposal realistic? In what ways would you criticize it if you were charged to do so?

3. At the time of this writing, the specific findings and recommendations of the FDA Nanotechnology Task Force have not been issued. Research what those findings and recommendations are. Do you think the recommendations are well supported by the findings or are there recommendations that you disagree with?

## 2.   Environmental Regulations

In 1928, Dr. Alexander Fleming was working at St. Mary's Hospital in London as a bacteriologist. One day, he returned to a plate culture of staphylococcus to find that it had been contaminated by a blue-green mold. What was most interesting about the contamination was that bacteria colonies next to the mold (*Penicillium*) were being dissolved. A subsequent effort by Fleming to grow the mold in pure culture resulted in the discovery of penicillin, a substance produced by the mold that had antibiotic properties.[21] Penicillin itself was not actually chemically isolated until World War II, earning Howard Florey and Ernst Chain the Nobel Prize in medicine because of its critical impact to the field.

The discovery of penicillin has rightly been hailed as a medical revolution. It was the first of a series of antibiotic drugs that transformed much of the way that medicine was practiced in the twentieth century. The very nature of the battle against infectious disease was radically altered, with the public quickly learning to rely on the availability of a pill to cure much disease. But the promise of these drugs was in some respects short-lived. The story is well known. The mechanisms of natural selection operated on the genetic structure of bacteria when they were exposed to antibiotics, causing them to mutate into strains that were resistant. Already in the 1950s—little more than ten years after the isolation of penicillin—it was apparent that tuberculosis bacteria had undergone mutations that provided it with resistance to streptomycin.

The situation has only gotten worse. Widespread use of antibiotics—many would more bluntly characterize it as negligent overuse—has resulted in the development of so many resistant bacterial strains that many antibiotics are now all but useless. And the condition is much worse in certain parts of the world where practices associated with the use of antibiotics are much less structured than others. A memoir published in 1996 by two senior epidemiologists of the Centers for Disease Control (CDC) noted how quickly local environments can be affected by antibiotic-driven bacterial mutation. At the time, the Pakistani city of Karachi had a population approaching 10 million and had many sectors without sewer lines and where waste would run in open ditches. The city tap water was unfit to drink and the general lack of sanitation was manifested by fecal material appearing in the food and water supplies, resulting in widespread infection by typhoid, salmonella, shigella, cholera, and other bacterial diseases. Blanket treatment with wide-spectrum antibiotics acted as a strong selection mechanism as bacteria were repeatedly cycled through human bodies, causing the development of resistant strains. It took only ten years for *Salmonella typhi*—the cause of typhoid fever—to develop from an organism that was highly sensitive to cheap and available antibiotics to being completely resistant to multiple, more costly antibiotics.[22]

In the United States, the CDC has become aggressive in promoting campaigns to educate both medical personnel and the public of the dangers of inappropriate antibiotic use. As a public-health organization, it describes antibiotic resistance as among its "top concerns." There is evidence that these campaigns are having some success in shaping public attitudes and providing an understanding that while antibiotics are indeed "wonder drugs," there are limits to how they can be used, and that inappropriate use has the risk of depriving future generations of their wonders.

### i.  Nanotechnology and the Environment

It is becoming clearer that nanotechnology has the potential to provide a similar kind of impact. The use of nanosized silver particles in many products has been promoted because of their antibacterial effects, making them useful in air fresheners, in shoe liners to destroy odors, and so on. But many environmentalists have voiced concerns that the resulting introduction of nanoscale silver particles into the environment at large may kill beneficial bacteria as well as other organisms. It is probably fair to say that the actual scope of the impact is as yet not fully understood, but that the danger of inadvertent degeneration of aspects of the environment is real. This concern about antibacterial effects of a certain aspect of nanotechnology resulted in a decision by the EPA to begin to regulate the use of nanoparticles for antimicrobial applications.[23]

Furthermore, this is just one of many potential ways that nanotechnology may impact the environment. A much more detailed discussion of the potential environmental effects of nanoscale particles is provided in the book in the Perspectives in Nanotechnology series about environmental impacts by Jo Anne Shatkin.

The decision of the EPA to begin to regulate even a single aspect of nanotechnology based on the potential for environmental effects represents a fundamental shift in policy and has the potential to be sweeping in scope. The EPA is one of the most powerful regulatory agencies in the United States. It currently employs nearly 20,000 people, having ten regional offices in addition to its headquarters office in Washington DC, as well as operating seventeen laboratories across the country. Formed in 1970 under the authority of President Nixon, the EPA is charged with implementing regulations that provide coordinated response to environmental concerns.

The legal basis for the various programs implemented by the EPA are provided by a suite of more than a dozen major statutes. Examples that are discussed in some detail include the Clean Water Act; the Safe Drinking Water Act; the Clear Air Act; the Federal Insecticide, Fungicide, and Rodenticide Act; the Toxic Substances Control Act; the Resource Conservation and Recovery Act; and the Comprehensive Environmental Response, Compensation, and Liability Act. But these acts should be considered merely as illustrations of the kind of regulatory schemes that may be used in controlling the environmental release of nanotechnology; other forms of legislation that provide authority to the EPA may also be used under different circumstances to achieve similar regulatory effects.

## a. Clean Water Act

The Clean Water Act (CWA) has its roots in the Federal Water Pollution Control Act Amendments of 1972. The basic function of the CWA is to provide a structure by which the discharge of pollutants into various waterways of the United States can be controlled. The structure provided by the CWA is organized around the concept of *"designated uses"* of waterbodies. The basic idea is that there are certain goals for the way in which a waterbody is to be used. Sometimes these goals are supportable by the waterbody, but other times are not. Another concept used by the CWA relates the conditions of the waterbody to the designated uses. Specifically, *"water quality criteria"* are defined as descriptions of those conditions that are necessary to support the designated uses. They are generally provided as quantitative measures, with examples including pollutant concentrations, temperature, pH, turbidity, and various toxicity measures.

The designated uses of any particular waterbody are intended to reflect a public determination of what generally desirable uses there are for the water. Some typical examples of use designations include use of the water for drinking, for recreation, for agriculture, as an industrial supply of water, and for the support of aquatic life. The water-quality criteria imposed on the waterbody by regulation are determined in accordance with these uses. It is not surprising, for instance, that water that is to be used for drinking may have significantly more severe constraints than water that is to be used as a source of industrial water. In some cases, further differentiation may be made within a particular use category. Recreational uses of water provides a good example because there may be wide variation in the level of human contact with the water in recreational settings. Operation of certain water-craft involves essentially no contact with the water, while activities like jet skiing result in limited contact and activities like snorkeling result in great contact with the water.

The CWA essentially provides regulatory authority to impose require-ments on water quality criteria in accordance with the determined uses. It is worth noting that a single waterbody may often be designated with multiple uses, reflecting the fact that it is viewed as having multiple purposes. Indeed, most waterbodies have multiple designated uses, requiring conformity with multiple water quality criteria. Since the water quality criteria associated with particular uses may involve different kinds of parameters, the overall set of criteria imposed on a given waterbody may actually address a diverse array of conditions.

*Characterization of Water Quality.* There are several different ways in which water quality criteria are characterized, including both quantitative and qualitative measures. In addition, the criteria may be applied to different por-tions of aquatic systems, including to properties of the water itself, to proper-ties of the biological species resident within the system, and to properties of inorganic structures that support the water such as bottom sediments, and the like. Different water quality criteria also seek to distinguish the tempo-ral effects of pollutant exposures, providing measures suitable for addressing acute short-term issues as well chronic long-term issues. Another aspect of water quality criteria is their relevance to different types of organisms, with different criteria being established for the protection of animals, plants, and microorganisms. Some such criteria take the form of population and diversity metrics that measure the quantity and variety of such organisms. Establish-ment of water quality criteria is expected to account only for scientific consid-erations, with economic and social impacts being inappropriate measures.

The manner in which the CWA might be applied to nanotechnology is thus diverse. Restrictions on the release of nanosized particles into waterbodies may be imposed directly, such as by limiting the volume of particles that have certain sizes that may be present within particular waterways. Such restrictions may vary according to the designated uses of the waterways, with water that is to be used for human consumption perhaps being subject to lower acceptable nanosized-particle densities. Such restrictions might also vary according to the composition of the particles, particularly as the different effects that nanosized particles have on different species become better understood.

Within this general framework, there are a number of specific issues that must be addressed for there to be any meaningful regulation of nanosized particles in waterways. Given the breadth of the definition of *"pollutant"* as used by the CWA, in which a pollutant may be essentially any material added to a waterway, there is essentially no doubt that such particles qualify as pollutants within the meaning of the act. Nevertheless, pollutants may only be regulated when there is a reasonable scientific basis to assert that the relevant water-quality parameters are adversely affected by the presence of such particles.

Currently, such bases are relatively poorly established, providing what is probably the greatest challenge in invoking the regulatory authority provided by the CWA to limit the level of nanosized particles in waterways. To establish meaningful water-quality standards relative to nanoparticle concentrations, the EPA will need to assemble a reasonable body of data establishing the kinds of harms that can result and defining risk levels for those harms.[24] It is fair to say that such studies remain in very early stages, including studies that evaluate toxicity in water environments, the way in which nanoparticles are transported through water environments, uptake mechanisms, and so on. Furthermore, such studies must meet high scientific standards before they can form a basis for regulatory action. They should, at a minimum, be peer-reviewed and published in reputable scientific journals. In performing such evaluations, it may be appropriate for the EPA to extrapolate known effects of larger particles, provided there is a legitimate scientific basis for doing so. For instance, data may exist on the effects of pollutants such as cadmium at larger scales, permitting extrapolation to determine the expected effect of such pollutants when they are provided in nanoparticle form.

*Control Mechanisms.* Parallel to these requirements of demonstrating risks, the EPA will need to have data related to control mechanisms. This may prove to be especially problematic and may represent a significant barrier to effective regulation. Specifically, it is not sufficient that some acceptable pollutant level be defined; there must be some demonstration that there are technically viable and economically reasonable mechanisms for polluters to limit discharge to the defined levels.

In many respects, this illustrates a more general characteristic of the state of regulation of nanotechnology at this point in time. Governments, including the U.S. government through the EPA, are largely in a stage of information gathering. This effort may be assisted by the provisions of the CWA, which grants authority to require those who produce pollutants to maintain records and make reports, as well as to perform inspections in the form of monitoring and sampling.[25]

This information-gathering process may integrate well with the permit function of the National Pollutant Discharge Elimination System (NPDES). This permit program is intended to control the level of pollution in waterways by regulation of so-called "point sources" of pollutant discharge. These are discrete conveyances into waterways, usually pipes from buildings but also potentially such structures as manmade ditches or other structures. While residential facilities are generally exempt from the program, most industrial facilities require a permit under the program if they provide discharges directly into a surface waterway. This permitting program allows more global management of pollutant discharges into waterways. While there are different types of permits and while each permit is specifically tailored to particular discharges, they are generally similar in that they specify effluent limits based on technology-based and/or water-quality-based standards.

It is as an adjunct to this permit program that much information about the discharge and effects of the release of nanosized particles into water environments may be gathered. Importantly, there is case law to the effect that this program provides the EPA with the ability to require all permit applicants to identify *all* toxic pollutants used or manufactured as intermediate or final byproducts.[26] The information gathering thus need not be limited to data regarding a facility's effluent actual discharge; because of the possibility that other toxic pollutants used by the facility *could* be discharged, the EPA has authority to collect information related to those pollutants. If a particular facility uses nanotechnology, the EPA thus has the ability as part of the permit process to inspect the nanotechnology operations of the facility, to collect records about the use and release of nanoparticles, and generally to perform those functions that allow it to obtain information that will be useful in defining regulatory standards related to nanotechnology.

More generally, the NPDES program provides a mechanism by which the effects of nanosized particles on water environments may be studied in a very direct way. It permits not only the collection of data defining levels of discharge, but also allows information to be developed on the effectiveness of various treatment technologies to achieve certain concentration levels in an economically viable manner. There has been some speculation that the manner in which nanoparticles are regulated under the NPDES program will mimic the pattern of regulation more generally under the NPDES. Initially, standards that were applied in granting permits tended to be technology-based and represented the best professional judgment of the permit writer. As better information was collected, the process of granting permits gradually became more standards based, with many permits now being issued only in accordance with the use of prescribed analytical methods, industry-specific effluent limit guidelines, and specific pollutant standards. This is likely to be the same pattern followed with regulation of nanosized particles, with permits initially being drawn according to the best professional judgment of the permit writer, but developing more specific water-quality-based criteria as more and better information is collected.[27]

It is perhaps worth commenting in addition on the ability of permits to be granted under "special conditions" that deal with unusual situations. The relative immaturity of understanding the environmental effects of nanotechnology likely qualifies as such a circumstance. What this allows is for a permit to be granted that requires the collection of "effects" data. Examples include ambient stream parameters, sediment, bioaccumulation in receptors, and the like. The designation of special conditions could also require the facility to perform certain tests that allow the accumulation of relevant information, including such things as toxic reduction evaluation studies and treatability studies. In cases where it remains practically impossible to establish precise limitations on effluent, the permit may designate certain management practices to be followed in regulating the discharge of nanosized particles. This is likely to occur only until the state of knowledge is sufficiently well developed to allow numerical effluent limitations to be specified.[28]

### b. Safe Drinking Water Act

In addition to the CWA, the Safe Drinking Water Act (SDWA) provides regulatory authority to the EPA to control the level of contaminants in water supplies, including the general authority to control levels of nanosized particles. While the CWA applies to waterbodies, the SDWA provides authority directed more specifically at drinking-water supplies. These regulations thus have the capacity to impact waterbodies that also come within the scope of the CWA, although there are many waterbodies subject to the CWA that are not sources of drinking water.

When it was initially passed in 1974, the SDWA approached the issue of ensuring safe drinking water by providing for treatment of water that was delivered to consumers, essentially focusing only on the end product. Amendments passed in 1986 and 1996 resulted in a change of focus. As the act is now implemented, there is a more balanced focus on preserving the quality of the drinking-water sources themselves in addition to providing for treatment of the water. But even so, compliance with the standards enforced through the act is evaluated by measuring the levels of defined contaminants in drinking water to ensure that they are below specified levels. Many water suppliers are required to prepare annual reports summarizing the level of contaminants in water.

*Risk Prioritization.* Application of the SDWA to nanotechnology is relatively straightforward since certain nanoscale structures will simply be classified as contaminants whose concentration in drinking water is limited. The approach for setting standards to limit such concentrations will follow the same protocols currently used in setting standards for other contaminants. This process requires the EPA to prioritize various contaminants according to the level of risk they present and their frequency of occurrence in water supplies. Incorporation of nanotechnology into this scheme thus requires determining how the level of risk presented by various nanoscale particles compares with the risk presented by other contaminants. In many instances, the nature of risk presented by different contaminants is different in character. For example, some contaminants are biological in nature, such as cryptosporidium microorganisms, while other contaminants are purely chemical in nature. A better understanding of the potential character of the risk presented by nanoscale structures may be found in the book in the Perspectives in Nanotechnology series on environmental impacts by Jo Anne Shatkin.

Once the risk prioritization has been established, the EPA sets a health goal that may reflect the different ways in which risk is manifested for different people, particularly for those groups that tend to have increased sensitivity to certain contaminants. This limit on concentration of the contaminant is set as much as possible to be consistent with the feasibility of achieving the determined health goal. In doing so, a cost-benefit analysis is often performed to balance the relative benefit achieved from limiting concentration amounts with the cost of compliance. These concentration limits, once established, may be enforced against any drinking-water supplier that fails to meet them. Administrative and punitive actions may be taken against utilities that fail to meet the established limits.

It is perhaps useful to consider a specific example of how the SDWA may be relevant to nanotechnology. Consider, for instance, a determination that silver particles having a characteristic dimension less than 100 nm present a risk of certain adverse health effects. An evaluation of the health risk presented by such particles results in a determination that the risk they present is relatively low, but that their ubiquitous use in certain commercial products has resulted in a generally high frequency of occurrence in water supplies. In evaluating the level of risk, an attempt is made to balance these somewhat contradictory considerations. In addition, suppose that it turns out to be relatively difficult to remove these particles from water supplies. Both the difficulty of removal and the relatively low health risk argue in favor of a fairly lenient standard. The level might be set to be near the current average level of occurrence to reflect a policy determination not to allow further increases in concentration beyond what have already become commonplace. Or the level might be set somewhat below the current average level to reflect a policy objective of reducing the prevalence of such particles in the water supply. In such cases, the level will most often be set so that the levels can be reduced without requiring the implementation of onerously burdensome procedures. This is consistent with the general determinations that the health risk is relatively low.

A completely different calculus might be used if the health risks are determined to be high. In such instances, much more drastic reductions in particle concentration may be viewed as necessary. Under those kinds of circumstances, the policy considerations that come into play are much more likely to view greater costs for compliance as acceptable to accommodate the increased health risks. None of this is particularly surprising. What the SDWA attempts to do is to provide a framework within which determinations may limit concentrations of potentially harmful substances in the water supply and provide an ability to enforce compliance with those limits.

*c. Clean Air Act*

The Clean Air Act (CAA) is implemented in a largely similar fashion. It provides for federally mandated limits on the air concentration of pollutants. A particular focus of the CAA is on specific air pollutants that are identified as *"criteria"* pollutants, meaning that their concentration limits are defined in terms of science-based criteria in a manner similar to the implementation of the SDWA. These science-based criteria may be developed to protect human health, and are referred to as *"primary standard"* criteria; or they may be developed to prevent damage to the physical environment of the planet, and are referred to as *"secondary standard"* criteria.

Current criteria pollutants identified by the EPA include ozone, carbon monoxide, nitrogen dioxide, sulfur dioxide, PM-10 (particulate matter with particles having diameters between 2.5 and 10 μm), PM-2.5 (particulate matter with particles having diameters less than 2.5 μm), and lead. The air concentration of nanoscale particles is thus arguably already regulated by limits imposed using the CAA on particles having a size less than 2.5 μm. But this size range still encompasses particles that are much larger than nanoscale and is not particularly narrowly directed to smaller nanoscale particles. If sufficient concerns are established about risks to human health or to the environment from nanoscale particles, it is possible for more narrowly defined criteria to be established. Such a classification may then result in specific limits being placed on particular nanoscale particles and their presence in the air.

The CAA is an example of federal legislation that allows the EPA to set national limits on air concentrations of pollutants. It is perhaps an obvious point, but one still worth making, that such limits establish only a minimal standard. Within a federal system like that in the United States, there are still other levels of government that can establish more stringent criteria. For example, it is common for state governments to impose higher limits on the pollutant concentration than the level established by the EPA. Indeed, this ability may be an important mechanism for recognizing particular local concerns about certain pollutant concentrations.

These lower levels of government are also involved when specific areas are identified as already being polluted at unacceptable levels. A state that contains such an area is required to develop a *"state implementation plan"* that defines how the state will implement regulations to bring the area into compliance. If and when the federal government implements specific regulations targeting nanoscale particles as pollutants, such state implementation plans are likely to become important. There will undoubtedly be a variation in the level of nanoscale pollutants across the country, and certain areas will have nanoscale-pollutant concentrations above the level that is established. The states that contain those areas will be obligated to develop implementation plans to reduce the pollutant levels. The EPA retains the authority to approve or reject state implementation plans. If a particular plan is deemed by the EPA to be unacceptable, the EPA itself may move in to take over enforcement of the Clean Air Act in that state.

*Characterization of Air Pollutants.* Procedural methods for characterizing the emission of air pollutants are developed by the Emission Measurement Center. This center is part of the Emissions Monitoring and Analysis Division in the EPA Office of Air Quality Planning and Standards. The center serves a number of different roles that allow it to promulgate standards for emissions testing and monitoring. Importantly, the center serves as an intermediary body between those who are devising the regulations and those who are to be subject to them. As part of this role, the center provides a mechanism for defining testing protocols used in evaluating pollution control and regulatory compliance. Indeed, the center is recognized as having acquired significant technical expertise that is likely to be critical in defining methodologies for quantifying nanoparticulate pollutants.[29]

There are a number of recognized technical difficulties in quantifying the level of nanotech air pollutants. For example, PM rules that set limits on the quantity of particulate-matter particles having certain sizes face certain implementation challenges. The EPA itself demands only that technology capable of 50 percent efficiency in capturing PM-2.5 particles be used. But, even so, technological limitations are generally likely to become less acute over time, and there are currently nanoscanning mobility particle sizers available that count particles in the 0.003–0.15 μm range.[30] A further difficulty with such very small particles is their temporal transience. Because they are so small, the distribution of nanosized particles tends to change quite quickly over time, at a rate that complicates the ability to obtain accurate counts. All of these practical measurement challenges translate into difficulties in establishing standards and evaluating compliance with them.

*Bases for Regulation of Airborne Nanoparticles.* There are at least two distinct ways in which regulation of nanoparticles might be implemented. It is conceivable that such particles could be designated pursuant to the CAA as criteria pollutants, in which case they would most likely be regulated in terms of their mass per unit volume of air. This is the traditional manner in which criteria pollutants have been regulated, although there is no specific requirement in the CAA dictating the use of such a measure. Alternative bases of regulation that might be used include the adoption of standards based on the number of particles rather than on their mass. With current monitoring technology, this is unlikely to be a particularly practical standard. The number distribution of nanosized particles is highly variable, making attempts to count the number of particles prone to a number of pragmatic challenges.

If nanoparticles are designated to be criteria pollutants, one likely designation to be used is the PM designation, but with considerably smaller sizes. For instance, anything less than a PM-1.0 designation would correspond to particles having a diameter less than 1000 nm, with PM-0.1, PM-0.01, and even PM-0.001 designations being used to regulate the air concentration of particles having diameters less than 100 nm, less than 10 nm, or less than 1 nm. In actual practice, these designations may be relatively difficult to implement. In principle, all that is needed is to establish limits on the appropriate concentrations but in practice, there may be a variety of technological barriers to implementation.

In addition to supporting various methods of measuring air pollutants, the EPA provides a number of modeling resources such as numerical software for simulating air pollution. Like conventional measurement techniques, the particular characteristics of nanosized particles make these modeling techniques generally unreliable for simulating the properties of airborne nanoparticles. The result is that current techniques—both measurement techniques and modeling techniques—make the regulation of airborne nanoparticles relatively impractical, at least under the current regulatory structure.

*Transnational Issues.* One of the difficulties that is particularly acute with air pollution is the potential for pollutants to enter a regulated region from somewhere over which the regulatory authority has no control. While this is a potential difficulty with almost any form of pollution, the ready ease with which air carries pollutants makes it of particular concern. For instance, suppose that the United States imposes limits on the concentration of nanoscale pollutants, but that regulations in Canada or Mexico are either much weaker or are nonexistent. Northern or southern regions of the United States may then find their air polluted with unacceptably high levels of nanoscale particles, but have no direct regulatory authority over the entities responsible for the production of such particles.

This possibility is addressed in the CAA by providing for a mechanism to limit contributions to air pollution in foreign countries by the United States.[31] The mechanism is based on the idea of reciprocal treatment by such foreign countries. As long as those countries provide a similar mechanism for regulating the flow of air pollutants into the United States, they will benefit from the provisions of the CAA to limit pollutant flow into their own countries. What triggers action by the United States to prevent the flow of pollutants into neighboring countries is the generation of reports or studies by recognized international agencies that they present a danger to public health. Subject to the reciprocity provisions, any country affected by the pollutant flow is provided with an opportunity to participate in public hearings held to produce or modify relevant implementation plans. Those involved with the production of nanotechnology products that result in the release of nanoscale particles into the atmosphere are thus well advised to monitor the tone of studies generated by international agencies regarding the effects of such particles as health pollutants.

### d. Federal Insecticide, Fungicide, and Rodenticide Act

A common feature of the Clean Water Act, Safe Drinking Water Act, and Clean Air Act is their general applicability to providing regulatory control over the release of certain contaminants or pollutants into the environment, with each of the different acts being directed to a different portion of the environment. These acts provided the EPA with the ability to regulate the release of nanoscale particles in a very general way. Other forms of legislation focus on the way in which particular substances might be used rather than on the aspect of the environment in which they are released.

This is perhaps most easily illustrated with the Federal Insecticide, Fungicide, and Rodenticide Act (FIFRA), which provides the EPA with the authority to control various forms of pesticides. The act was substantially modified in 1996 with passage of the Food Quality Protection Act, which sought generally to provide more consistency to the regulatory scheme. In particular, the Food Quality Protection Act mandates a single standard for the level of pesticides that appear in foods. It also provides special provisions to protect children, in addition to more administrative functions like modifying the approval process for safer pesticides and creating certain incentives for developing crop maintenance tools.

The FIFRA is relevant to nanotechnology because of the promise that nanotechnology provides in the development of new types of pesticides. Already, products are being developed in which nanoscale droplets are added to pesticides to prevent the separation of components. The incorporation of nanotechnology into pesticide products is also being pursued in the form of delivery systems that make use of nanostructures to reduce solvent content, and enhance certain physical effects of the pesticides such as dispersity, wettability, and droplet penetration strength. There are, moreover, proposals in which dynamic nanotechnology systems are used to diminish pesticide runoff after application. In addition, there may be more direct uses of nanostructures as active pesticides themselves, extending the now largely conventional uses of nanosized silver particles as antimicrobial agents.

*Pesticide Registration.* The FIFRA operates generally by establishing a registration requirement for pesticides after data have been collected and analyzed to define appropriate dosages and risks consistent with a certain effectiveness. Because of the broad sweep of the FIFRA to regulate pesticides, there is no question that such substances that incorporate nanotechnology are subject to its provisions. Where questions arise is when an existing conventional pesticide is modified to incorporate nanotechnology. Conventional pesticides are subject to a registration requirement and the issue attempts to resolve whether the incorporation of nanotechnology into a registered pesticide so changes the character of the pesticide that a new registration is needed.

The most likely result seems to be that the prior registration would be insufficient.[32] The fundamental factors to be balanced in determining whether to require a further registration are the claims that are made regarding its efficacy and its composition. The general expectation is that the incorporation of nanotechnology into an existing pesticide will sufficiently affect the balancing of these factors that a further investigation and registration will be needed. One example that has been advanced relates to the antimicrobial properties of nanoscale silver particles.[33] Because of these properties, the incorporation of silver nanoparticles into pesticides is something that is sometimes considered. And the relatively good effectiveness of such particles may greatly affect the claims made for pesticides that incorporate them.

What this basically accomplishes is to recognize that the inclusion of nano-scale materials in pesticides results in a different balancing of risks and benefits than does a similar conventionally made product. An evaluation of how these risks and benefits interplay is key in making registration decisions. At the moment, it is fair to say that neither the particular risks posed by nanomaterials nor the particular benefits they can provide are accommodated by the registration guidelines. This situation is very likely to change as registration requests are increasingly received by the EPA for pesticides that incorporate nanotechnology. As least the part of the registration process that involves the collection of relevant information provides an effective mechanism for obtaining the sort of information that can then be incorporated into the registration guidelines to improve the way products containing nanomaterials are regulated.

### e. Toxic Substances Control Act

Many of the issues that arise under the FIFRA also relate to the application of the Toxic Substances Control Act (TSCA). This act takes the approach of regulating *"chemical substances,"* which is defined to be "any organic or inorganic substance of a particular molecular identity, including—(i) any combination of such substances occurring in whole or in part as a result of a chemical reaction or occurring in nature and (ii) any element or uncombined radical."[34] With such a broad definition, there is no serious question than nanomaterials are chemical substances within the meaning of the act.[35]

There are two principal ways in which the TSCA may be applied. First, there are provisions that authorize the direct regulation of various activities related to chemical substances, including their manufacture, processing, distribution, use, and disposal. Whenever a chemical substance potentially presents an "unreasonable risk" of injury to health or the environment, the EPA is granted a number of regulatory powers. These principally include the ability to limit or prohibit those activities subject to regulation, the ability to require warning and instruction labels, the ability to require monitoring and testing, and the ability to define disposal procedures.[36] Any of these powers has the potential to be applied to products that incorporate nanotechnology and are likely to be invoked, particularly in the relatively early stages of their use when risks are not especially well defined.

In addition to such direct regulatory authority, the TSCA provides the EPA with the power to require testing of chemical substances under a number of different circumstances.[37] The power certainly exists whenever it determines that the substance might present an unreasonable risk to health or the environment. But it also exists when a substance that is produced in "substantial" quantities has the potential to be used in a way that provides "substantial" exposure to humans or the environment, if there are insufficient existing data to determine its full health and environmental effects. This is certainly a potential basis under which the EPA might require a variety of tests of products that incorporate nanomaterials. In many respects the conditions under which the direct regulatory powers arise will also implicate the testing powers. Many nanotechnology products are actively being introduced in ways that result in considerable exposure to humans and the environment at a time when many critics question the state of knowledge of their safety. The TSCA provides an important regulatory tool to exert some level of governmental control on how such products are introduced into the marketplace and how they should be tested to ensure their safety.

The TSCA also includes a provision that may be used to initiate recall procedures of nanotechnology products under certain circumstances. This is accomplished by authority that is granted to initiate a civil action for seizure of "imminently hazardous" substances.[38] The authority that is granted is actually quite a bit broader than merely the ability to initiate recalls. More generally, a civil action may be initiated to require notifications to individual purchasers of risks associated with the products, to require public notifications, or to require replacement or repurchase of products, in addition to recalls. This power can potentially be applied when a basis is discovered to believe that nanotechnology products are imminently hazardous.

One important area in which the TSCA differs from the FIFRA is in the level of regulation of research and development activities. The TSCA is considerably less restrictive in the requirements it places on such activities than is the FIFRA, resulting in much greater potential for control to be provided over the development of pesticides that contain nanomaterials than over other types of products. Specifically, the TSCA requires only that known hazards be identified to the EPA, the amount of material to be produced be reasonably necessary for the research, records be maintained, and the research be supervised by a qualified person.[39] This is in marked contrast to the much more stringent permitting requirements imposed by the FIFRA on pesticides, even during periods of research development.

### f. Resource Conservation and Recovery Act

Another important area in which the EPA promulgates regulations deals with the disposal of waste. This is, in fact, a critically important function of the activities of the EPA, in many ways more so that its regulation of products that are delivered or sold to consumers. Waste products have the potential to have substantial environmental impacts through a number of different pathways, examples of which include migration through groundwater, adherence to soil, and transport via waterways or through air. Any of these mechanisms has the potential to allow the waste to be consumed in some fashion by biological organisms. In certain respects, nanoscale materials present a number of unique regulatory difficulties: the size of the materials makes them relatively difficult to control and the relatively large total surface area of groups of particles has the potential to increase their toxicity.

While some of the acts discussed earlier have the potential to impose severe limits on the disposal of waste containing nanomaterials, the Resource Conservation and Recovery Act (RCRA) provides more targeted authority to regulate the disposal of waste materials directly. The approach of this act is not to focus as much on the environmental structure in the way the CAA or

CWA might, but instead to focus on the generation, transportation, management, and disposal of waste.

Generally, the RCRA provides expansive power to the EPA to control virtually all aspects of "hazardous" wastes in the United States. There is essentially no question that the breadth of this power enables the EPA to regulate the management and disposal of waste that includes nanomaterials. In a manner similar to the exercise of regulatory authority under the other acts described in this chapter, this power may be exercised either by using existing or by promulgating new regulations. The use of existing regulations is appropriate when nanomaterials are incorporated into waste products that are already subject to regulation. And the promulgation of new regulations is appropriate when nanomaterials are formed in new kinds of waste.

The power to define new regulations provides flexibility on the part of the EPA to address hazards arising from nanomaterials that may as yet not be well understood. At the moment, there are four types of hazards that have been defined by the EPA: ignitability,[40] corrosivity,[41] reactivity,[42] and toxicity.[43] It is conceivable that the impact of nanomaterial waste on the environment might involve some hazard that does not fall easily within the scope of these defined hazards, which were developed with more conventional materials in mind. As such new risks become more apparent, this authority may become of greater importance in defining the scope of regulatory involvement by the EPA in managing nanotech waste.

Even within the existing framework for defining hazards, there are a number of waste products that have been specifically designated by the EPA to be hazardous. These specific identifications are organized into four "lists": the F list, which identifies waste products from various industrial processes that may arise in different industry sectors ("nonspecific source wastes");[44] the K list, which identifies waste products from specific industries;[45] and the P and U lists, which identify specific chemical products that become hazardous waste when discarded.[46] These lists are not intended to be exhaustive and there are many waste products that fall within one of the four defined hazards that are not included on any of the lists.

None of the lists currently includes wastes defined in terms of nanoscale materials. But many of the various industrial activities underlying the structure of both the F and K lists are likely to either use nanoscale materials or to produce them as by-products. For instance, the K list includes wastes from such industries as the explosive industry, the pesticide industry, and the ink-formulation industry, all of which are currently increasing the use of nanomaterials in their products.

*When is Hazardous Waste not a Hazard?* It is worth noting that at least some commentators have remarked on the possibility that the current framework for regulating hazardous waste has the potential to produce unintended consequences.[47] There is the possibility that certain formulations may be inadvisably identified as hazardous. Consider, for instance, the case where a relatively large mixture of wastes includes some small quantities of materials that are on one of the waste lists. The entire mixture may then be designated as hazardous waste. Perhaps even more important, though, is the fact that nanomaterials may exhibit significantly different properties even when they are similar chemically to bulk materials. This leads to the risk that a nanoscale formulation of a substance will be designated as hazardous because of a characteristic of the bulk material that is not shared by the nanomaterial. These have the potential to inhibit the development of certain nanoscale materials that might provide not only a much reduced environmental impact but potentially have a positive effect on the environment. This may remain as an issue that needs to be addressed.

*g.   Comprehensive Environmental Response, Compensation, and Liability Act*

The various acts described so far are focused in certain specific respects. The Comprehensive Environmental Response, Compensation, and Liability Act (CERCLA) takes a different approach by providing wide environmental regulatory authority that may potentially be invoked when one of the more specific acts does not apply. CERCLA is commonly referred to as *Superfund* and was enacted in 1980 on the tenth anniversary of Earth Day; it was formed primarily to address cleanup of hazardous waste sites. But it was fashioned in such a broad fashion that it finds wide applicability in filling regulatory gaps that might exist with other statutes.

The basic administration of CERCLA identifies sites contaminated with hazardous waste and quantifies the risk level associated with the hazard. The sites that are identified are prioritized on a National Priority List to manage a schedule under which they are investigated in detail and subject to clean-up activity. *"Hazardous substances"* are defined under this act in the broadest possible terms, making reference to other acts[48]—some of which have been described earlier—and providing omnibus authority to identify any substances that "when released into the environment may present substantial danger to the public health or welfare of the environment."[49] With such a broad definition of hazardous substances, there is no question that nanomaterials are potentially subject to the act.

*Retroactive Effects.* An interesting aspect of regulation under CERCLA is that a substance that is designated as being "hazardous" enjoys no exception because of its status at the time of its designation. Irrespective of whether the substance is one that is merely in a state of development or has been widely released, the designation as hazardous causes it to be subject to the regulatory provisions. The rationale for this approach is not to punish those who might have released a product that may cause harm but instead to provide a mechanism for recognizing that mistakes may have been made in the past and can now be corrected. There is accordingly a certain level of risk faced by those who are bringing nanotechnology products to market. Should those products be found to be "hazardous" within the (potentially very broad) meaning of Superfund, they will be subject to regulatory provisions that can profoundly affect their market.

The structure of CERCLA is generally focused on the cleanup of "sites" where there has been a release of hazardous material. In this respect, it is somewhat different from the regulatory provisions of other environmental acts that focus more broadly on general releases of materials into the environment. This structure results in there being two distinct levels of risk assessment that are undertaken in order to trigger some remedial action.

First, there is a threshold evaluation of whether a particular material poses sufficient risk that it should be regulated. CERCLA itself already contemplates making this evaluation in concert with similar risk evaluations performed under the TSCA or FIFRA. To the extent nanomaterials are found to be of concern under those acts, they are likely also to be of concern under CERCLA. Second, a more specific evaluation is performed of whether a particular site should be subject to some kind of remedial action. There are a variety of mechanisms for making such a determination, including the use of the National Contingency Plan, the National Priorities List, the Hazard Ranking System, and various state-level programs.

*Remedial Responses.* Once a determination has been made to take some action, there are a number of different forms that such action can potentially take. For example, the EPA has the power to finance and oversee the removal of the hazardous material and other remedial actions,[50] to compel third parties to disclose information,[51] to acquire property needed to implement its remedial action,[52] to fund response actions and peripheral matters,[53] and to conduct brownfields evaluations.[54] The term *"brownfield"* refers generally to land that was previously used for industrial or commercial purposes that resulted in contamination of the land, but which retains the potential to be reused after appropriate cleanup. These various powers have been shown over time to be flexible, permitting the EPA to tailor its cleanup approaches in different ways depending on the specific nature of the contamination. There is every reason to expect the same flexibility to be available when sites are found to be contaminated with some form of hazardous nanomaterial.

## h.  Assessment

The various provisions discussed are merely some of the many regulatory acts that provide power to the EPA to prevent environmental contamination and to respond when contamination occurs. All of these acts have the potential to be applied to nanotechnology with the same force with which they have been applied to other types of hazards. What is perhaps interesting about the array of regulatory provisions is that they take a number of different approaches, which the selection of specific acts discussed here has attempted to highlight. For example, some provisions focus on regulations that are pertinent to particular sectors of the environment—the air (Clean Air Act) or the water (Clean Water Act or Safe Drinking Water Act), for example. Other provisions focus on the nature of the vehicle by which a hazardous substance might be delivered to any sector of the environment; the Federal Insecticide, Fungicide, and Rodenticide Act focuses specifically on forms of pesticide. This is a relatively specific focus on a delivery mechanism. The Toxic Substances Control Act is similar in focusing on the delivery vehicle of hazardous substances, but does so in a more general way by regulating the release of any chemical substance that meets the defined requirements. The Resource Conservation and Recovery Act can fairly be described as inter-mediate in its approach. In a sense, it is specific since it focuses on delivery mechanisms that involve the disposal of waste, while at the same time being broad by not otherwise limiting the nature of the waste. The Comprehensive Environmental Response, Compensation, and Liability Act takes yet another approach by focusing on specific sites, providing a mechanism for address-ing environmental contamination in any environmental sector of that site and arising from any mechanism.

What the resulting combination of regulatory acts produces is a scheme that can readily address potential environmental hazards produced by nanotechnology using the approach that turns out to be best suited for a par-ticular type of contamination. This is potentially very useful and is remark-able in that such flexibility is provided with a framework that was designed without nanotechnology in mind.

But what this section has not really addressed—and which has the poten-tial to be equally or more important—is the ability of nanomaterials to have a positive impact on the environment. Nanomaterials may be used effec-tively as remediation tools that can help mitigate the risks associated with many more conventional kinds of hazardous substances. One example that has been noted is the ability of many nanomaterials to promote the degrada-tion of chlorinated hydrocarbons—a class of substances included in conven-tional pesticides like lindane and DDT, as well as in dioxin and furan waste products—that are known to accumulate in food chains and cause a num-ber of disorders in human beings and animals. There is also evidence that nanosized particles of iron may be useful in cleaning up carbon tetrachlo-ride contamination in groundwater. The great likelihood is that many more discoveries are yet to be made of how nanotechnology can be beneficial to the environment rather than causing harm. An interesting question is how these potential benefits may be most effectively accommodated by whatever regulatory scheme results.

### Discussion

1. A variety of different specific regulatory acts have been discussed in the main text and the way in which they may be used to regu-late nanotechnology in environmental contexts. What gaps can you identify in the coverage provided by the acts? That is, are there any activities that you believe are of concern and for which there is cur-rently no adequate regulatory authority? What specific proposals can you make to fill these gaps—do you envisage modifications to existing regulatory authority or are the gaps so large that a new kind of authority is needed to address the concerns?

2. Research the provisions of the National Pollutant Discharge Elimi-nation System program implemented in the United States. The EPA is subject to a mandate under the CWA to create control strategies

for toxic pollutants under this program. What challenges do you see confronting the EPA in attempting to meet this mandate?

3. The SDWA includes provisions that allow states to grant variances from standards for systems that serve relatively small numbers of people when there are financial difficulties associated with full compliance. For example, systems that serve up to 3300 people may be granted by the state a variance if it is too costly to comply and certain approved variance technology is installed. Systems that serve 3301–10,000 people may be granted a variance with approval of the federal EPA. While the terms of the variances are required to ensure no unreasonable risk to public health, do you see any issues specific to nanotechnology that cause you to be concerned about the availability of such variances?

4. One major exception to the regulatory authority of the EPA to regulate the disposal of waste under the RCRA relates to household wastes. The definition of *"hazardous waste"* specifically excludes "household waste."[55] A significant source of the release of nanomaterials into the environment is likely to be through the disposal of many commercial household products that include nanomaterials. How do you think this exception will affect the overall success of regulatory programs governing the disposal of waste that includes nanomaterials? How can this be corrected? Research other exemptions that exist to the definition of hazardous waste (see 40 C.F.R. §261.4). Can you see any other exemptions that have the potential to act counter to the overall goals of the RCRA as applied to nanotechnology?

5. There are a number of different ways in which used materials have been treated. The discussion of the RCRA has focused primarily on disposal. But recycling of consumer products is also a viable—and perhaps increasingly desirable—technique for treating used products. What potential issues do you see with trying to apply the regulatory scheme set forth by the RCRA to recycling of products that incorporate nanotechnology? One aspect you may wish to address is that recycling may offer different pathways for nanomaterials into the environment that are not available when conventional disposal techniques are used.

6. Consider the life cycle of each of the following products that may contain nanomaterials, including the creation of the product, its use, and its disposal. Which of the regulatory acts described in this section have provisions relevant to the product? Discuss how they might apply. What conclusions can you draw from this exercise about the effectiveness of environmental regulations in their current state to nanotechnology?

   a. Trousers treated with a nanomaterial stain-resistant coating

   b. A wound dressing for burn victims treated with nanoscale silver material as an antimicrobial agent

    c.  A tennis racquet made using carbon nanotubes

    d.  An antiwrinkle cream made using nanosomes

    e.  A windshield film made using chemical self-assembly of nano-materials to prevent sticking

    f.  A computer monitor that uses nanotubes to stimulate individual pixels

    g.  Nanoparticle powders or nanoparticle dispersions produced for use by researchers

7. The structure of CERCLA is such that its provisions are triggered by the release of a substance that is "hazardous" at least at the level of a "reportable quantity" into the environment. The underlying assumption of this structure is that larger quantities of hazardous substances pose greater risks. Is this assumption necessarily valid for nanomaterials? If not, how would you propose to modify CERCLA to account for the unique concerns posed by nanomaterials?

8. In addition to providing for coordination among different acts administered by the EPA, CERCLA provides for coordination between the EPA and the U.S. Agency for Toxic Substances and Disease Registry. Research the role of this agency in regulatory matters. How might the mandate of this agency affect the regulation of nanotechnology?

9. Much of the discussion of environmental regulatory issues of nanotechnology is speculative because many of the specific environmental impacts are not yet well understood. To what extent is there a danger that regulatory provisions will be invoked too soon, resulting in regulation of nanotechnology that is inappropriately restrictive? To what extent is there a danger that the application of regulation of nanotechnology in an environmental context will be delayed too long, resulting in significant harms that will be difficult and costly to remedy? How should the current level of ignorance about certain aspects of the potential impact of nanotechnology be balanced by those in a position to invoke regulatory authority?

10. At the end of this section, the potential for nanotechnology to have a positive effect on the environment is raised. This positive effect may be not only in the ability of nanotechnology to provide alternatives to more environmentally damaging conventional substances, but to be used in an active way for remedial purposes. At the same time, the bulk of this section has noted the various ways in which existing regulatory provisions may be invoked to inhibit certain uses of nanotechnology. Do you think there is a danger that these provisions may be used to inhibit even actively positive uses of nanotechnology on the environment? What factors do you think might affect whether environmental regulation is used in such a counterproductive fashion? To what extent is public perception of the dangers of nanotechnology likely to be a factor?

## 3. Regulation of Exports

In 1986, a Guyanese student, known to the world only as "C.S.," met a Soviet physicist working for the United Nations at a subway station in the Queen's borough of New York. C.S. had known the physicist, Gennadi Zakharov, for about three years. He had been helped by Zakharov in his studies and in securing a job with a defense subcontractor. During their encounter in the subway station, C.S. handed over an envelope that contained classified documents describing United States Air Force jet engines in exchange for $1000 in cash. Immediately after the exchange, agents of the Federal Bureau of Investigation swarmed around Zakharov, subduing him and shackling him in handcuffs. It was only then that Zakharov discovered that C.S. was working undercover with the FBI and that his plan to steal secret technical information from the United States had failed.[56]

The Zakharov incident is one of only a handful of espionage cases that are known in which a scientist has been caught attempting to steal secret government information. Either scientists are unusually capable spies or there is some cultural impediment that discourages them from participating in this kind of activity. In a certain respect, the notion that scientists would refrain from partaking in government espionage is surprising because there is undoubtedly a philosophical predisposition among them to favor the open exchange of information. Scientific research is undoubtedly one of the most international of activities, with scientists routinely acknowledging the value of exchanging ideas and information regardless of where those ideas originate.

But the reality is that the world is structured with political boundaries. These boundaries define countries that have different ideologies and compete with each other, each believing that its own worldview is philosophically superior. And the leaders of these nations routinely take the position that scientific knowledge has the potential to offer them a competitive advantage that will enhance their ability to protect and promote their own values. Nations accordingly impose restrictions on the dissemination of scientific and other technical knowledge to those outside their own borders.

### i. Classification

The formal classification of information provides the most restrictive form of control. Information that has been formally classified by a government not only limits the people to whom it can be disclosed, but also generally requires that those individuals have met certain standards of trustworthiness after an extensive security investigation. While every country has its own classification system, a common characteristic of these systems is that there is a hierarchy of classification levels. Information is assigned to one of the levels and only those who have a clearance level at or above that level are permitted to have access to the information.[57]

For example, there are currently three classification levels used in the United States and these are correlated with the potential damage that could be caused to national security. The lowest classification level is *"confidential"* and refers to information that could cause "damage"; the intermediate classification level is *"secret"* and refers to information that could cause "serious damage"; and the highest classification level is *"top secret"* and refers to information that could cause "exceptionally grave damage."[58] At times, there has been a fourth (lower) level, *"restricted,"* that referred to information that could cause "undesirable effects." Similar kinds of classification levels exist in virtually every classification system used by every country in the world. An important point to be made is that the government official who designates information as classified is required to describe with some specificity what danger is posed by release of the information. He must also estimate a date when the information can be declassified, although this can of course be reevaluated as circumstances dictate.

It is not surprising that the number of classification levels is different now than it was previously—and is likely different than it soon will be. In the United States, classification of information is a function of the executive branch and is therefore implemented with the authority of the president rather than by the legislature. It is almost customary whenever the party in control of the executive branch changes for there to be modifications made to the classification system.

There is no serious doubt that there is currently information related to nanotechnology that is subject to these classification restrictions. While attempting to identify what topics are classified is an exercise in pure speculation, obvious candidates include nanotechnology that has military and surveillance applications. But even much more mundane research may also be classified because of its potential to lead to applications that are considered to be sensitive.

Researchers who work with classified nanotechnology information have a responsibility to protect and control access to the information so that there is no unauthorized disclosure. The level of security applied may vary depending on the sensitivity of the information, but typical requirements include storing the information in an appropriately secure safe, shredding documents when they are to be discarded, and refraining from working on classified projects outside of a designated work area.

The penalties for disclosing classified information can be severe. Any government employee who merely removes classified information to keep it at an unauthorized location may be subject to a fine of up to $1000 and a year of imprisonment. If that employee instead transmits the information to someone thought to be an agent of a foreign government, both the potential fine and the potential term of imprisonment increase by a factor of ten. Penalties can also be imposed under some circumstances when there is a violation of the Espionage Act.[59] The Espionage Act generally attempts to prevent information related to "national defense" from being obtained by foreign powers.[60] It is even possible under this act for a sentence of death to be imposed if defense information is conveyed to a foreign power that concerns such things as "military spacecraft or satellites [and] early warning systems,"[61] all of which will increasingly incorporate nanotechnology-based structures.

The potential for nanotechnology to be applied in a manner that directly implicates the security concerns of governments is substantial. But actual classification of nanotechnology-related information is not the only way in which the dissemination of such information is controlled. There are a number of other laws that impact dissemination even of information not subject to formal classification. Foremost among these are the Export Administration Regulations (EAR) promulgated by the Department of Commerce and the International Traffic in Arms Regulations (ITAR) promulgated by the Department of State. These regulations are of particular concern to nanotechnology researchers because such research is, like much fundamental scientific research, international in character. What they prohibit is the "export" of certain types of information. This is a much broader concept than the transmittal of information to a foreign power and can severely limit the kinds of interactions that researchers take for granted, particularly in academic settings. Also of some relevance may be boycotts imposed against specific countries by the Office of Foreign Assets Control (OFAC), which is part of the Treasury Department.

## ii.  Export Administration Regulations

An *export* covers any action that transfers information or a physical item to a foreign country. Critically, this includes transfers to foreign nationals, wherever they may be. The disclosure of controlled information to a French citizen who is studying in the United States on a properly granted visa may be a violation of these export-control regulations. This disclosure need not be in writing and can take the form of an oral discussion, a demonstration of some nanotechnology device, and posting information on an Internet site in a way that makes it accessible to non-U.S. persons, among other types of disclosures.

The Export Administration Regulations are governed by the Export Administration Act of 1979.[62] This act sets forth a rather complex structure for determining whether any particular item or piece of information can be exported without first obtaining permission of the government in the form of a license. A rough overview of the considerations that are made is provided with a flow diagram in Figure II.2.[63] In considering this flowchart, though, caution should be exercised because it glosses over some obscure details that could prove relevant in determining whether a license is required.

The flowchart brings out a number of issues that the structure of the Export Administration Regulations seeks to take account of in limiting the disclosure of objects or information. Perhaps most fundamental is the fact that the EAR is not all-encompassing. Only certain kinds of objects and information are "subject to the EAR." This is checked at Step 1 in the flowchart. For example, the control of certain kinds of information and objects might be

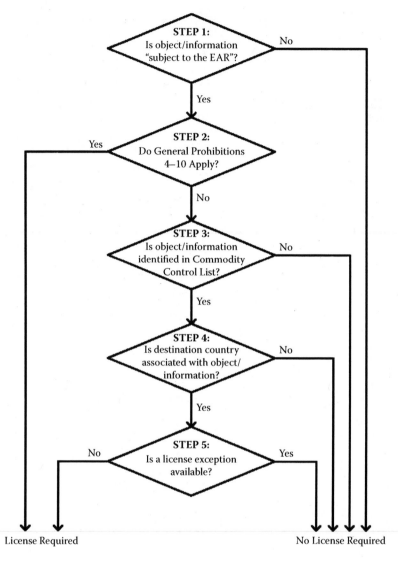

**FIGURE II.2**
Flowchart summarizing how to determine whether an object or information requires an export license.

assigned to the exclusive jurisdiction of some other agency, or they might be sufficiently public that it does not make sense for them to be restricted. Activities that are subject to the EAR include those related to proliferating nuclear, chemical, and biological weapons—activities that could potentially have a nanotechnology component—as well as certain encryption activities.[64] Other items and activities that are subject to the EAR are those that are in the United States, are of U.S. origin, or have some sufficient U.S. component, but are not in public circulation.[65]

Technology that is subject to the EAR in this way is further defined by a Commerce Control List that defines fairly specifically what forms of technology are restricted. If technology subject to the EAR is not to be used for certain "general prohibitions," then this list needs to be checked to ascertain the precise scope of any restrictions. The flowchart therefore shows a check at Step 2 for whether any relevant general prohibitions apply, with a further check at Step 3 for whether the specific object or information is identified in the Commerce Control List.

While there are ten general prohibitions, only seven automatically result in the need for a license. These seven reflect relatively serious concerns: engaging in actions specifically prohibited by a denial order (Prohibition 4), exported to specifically prohibited end uses or end users (Prohibition 5), exporting to embargoed destinations (Prohibition 6), supporting proliferation activities (Prohibition 7), shipments involving vessels or aircraft (Prohibition 8), violating terms of an existing license (Prohibition 9), or proceeding in the face of knowing about a violation (Prohibition 10).

The remaining prohibitions may trigger the need for a license depending on the specific content of the Commerce Control List. This list is quite extensive, with the scope of technologies specifically identified being extremely diverse. It is structurally organized into ten "categories," multiple of which have the potential to encompass nanotechnology:

0. Nuclear Materials, Facilities and Equipment and Miscellaneous
1. Materials, Chemicals, "Microorganisms," and Toxins
2. Materials Processing
3. Electronics
4. Computers
5. Telecommunications and Information Security
6. Lasers and Sensors
7. Navigation and Avionics
8. Marine
9. Propulsion Systems, Space Vehicles and Related Equipment

Substructure within each category defines five groups:

A. Equipment, Assemblies and Components
B. Test, Inspection and Production Equipment
C. Materials
D. Software
E. Technology

Individual entries within each group have a number of different pieces of information, including a specified reason why the technology is subject to control, a list of countries to which export is controlled, and an identification of specific items that are subject to control. Currently, there are fourteen possible reasons for control of export, many of which it is easy to envisage implicating nanotechnology:

AT Antiterrorism

CB Chemical and Biological Weapons

CC Crime Control

CW Chemical Weapons Convention

EI Encryption Items

FC Firearms Convention

MT Missile Technology

NS National Security

NP Nuclear Nonproliferation

RS Regional Stability

SS Short Supply

UN United Nations Embargo

SI Significant Items

SL Surreptitious Listening

If an object or information is identified in the list, then further investigation is needed at Step 4. The fact that technology is subject to control does not amount to a blanket prohibition on exporting the technology. Instead, each item in the Commerce Control List is associated with one or more countries that are of concern. The check made at Step 4 is whether the destination country is one of those identified. If not, it is permissible to engage in the export without a license. But if the country is identified, then licensing requirements are specified in accordance with the reason why the item is controlled.

The final step in the flowchart, Step 5, is a check to see whether some exception to the licensing requirement might exist. Such exceptions may apply whenever one of the other three general prohibitions applies: exports to certain countries (Prohibition 1), exports from abroad of items incorporating more than minimal U.S. content (Prohibition 2), and exports from abroad of items produced from U.S. technology (Prohibition 3). There are a number of different license exceptions and the manner in which they apply may differ depending on the item to be exported.

For instance, one exception is provided for "shipments of limited value." The Commerce Control List includes a specification of the threshold value for items that are eligible for this exception. Another exception provides for an exception when the items are destined for "civil" end users (as contrasted with "military" end users). A variety of temporary exports are eligible for a license exception, as is beta test software that is intended for public distribution. Exports that are made for humanitarian reasons are also eligible for an exception. Not surprisingly, many of the exceptions reflect a preferential posture with respect to some countries and not with others.

This overview of how to determine whether a license is required for the export of a particular technology is necessarily superficial. The regulations governing exports are complex and it requires careful consideration of the precise nature of the restrictions imposed on particular items before there is any export. Fundamentally, what the Export Administration Regulations seek to do is to implement policy considerations that take account of five basic factors: what the item is, where the item is going, who will receive the item, what the recipient will do with it, and what else the recipient does.[66]

### iii.  *Fundamental Research Exclusion*

One additional consideration that is of particular relevance to nanotechnology bears on at least two of these factors—what the technology is and what the recipient will do with it. There is a very large exclusion to the Export Administration Regulations for *"fundamental research,"* which the regulations define as "basic and applied research in science and engineering, where the resulting information is ordinarily published and shared broadly within the scientific community."[67] This exclusion is among the strongest legal recognitions that basic research is conventionally performed in an atmosphere of openness. It provides one of the most important exceptions for excluding aspects of nanotechnology that would otherwise be subject to control. Nanotechnology remains at a stage of development where many of the most important advantages are still the result of such fundamental research.

Research that is conducted at universities is the most obvious candidate for qualifying as fundamental. Indeed, the regulations themselves acknowledge that research conducted at universities "normally" meets the requirements.[68] What causes university-based research sometimes not to qualify is the acceptance of restrictions on publication of information that results from the research. Such restrictions are generally inconsistent with the broader definition of fundamental research that the exception attempts to embrace. Frequently, these kinds of restrictions are imposed by industrial sponsors of the research. The nature of industrial interests is such that the research they sponsor is often the most difficult to classify as "fundamental." Because corporate research is generally performed to obtain some competitive advantage, it is much less likely to be disclosed than university research. The restrictions on public disclosure associated with the ability to obtain a patent also act more strongly to suppress the results of corporate research because many more patent applications are filed by corporate interests than by university interests.

An important mitigating factor to these limitations on use of the exclusion—for both corporate and university research—is that research that is not "fundamental" because its disclosure is restricted has the potential to become "fundamental" when such restrictions are lifted or expire naturally. This exception thus appears in some ways to be easy in application. It simply considers whether any disclosure restrictions are in place at the time of export. In principle, a nanotechnology researcher who is working on a project that produces results whose export is restricted by the EAR merely has to determine whether any disclosure restrictions are in place when she wishes to perform the export. If there are, there is no question that she is not permitted to perform the export; if there are no such limits, the fundamental-research exception may apply.

### iv.  International Traffic in Arms Regulations

But some commentators have suggested that the exception as it is applied in a regulatory context is not as strong as it appears in the abstract.[69] In certain measure, this is due to the way in which the ITAR regulations are implemented. This is the other of the two major sets of regulations that provides restrictions on exports of technical knowledge and both of them treat the export of fundamental research in a similar manner. The ITAR is more specifically directed at military technology than is the EAR. It controls the export of "defense articles" and "defense services," the definitions of which make reference to a "munitions list." The munitions list includes about twenty categories of munitions, all of which have the potential to incorporate nanotechnology, although some categories lend themselves more naturally to the use of nanotechnology than others. For instance, "directed energy weapons" and "toxicological agents" are highly likely to incorporate nanotechnology, but any of the categories might easily include munitions developed to use NEMS structures, nanoenergetic explosives, coatings having embedded nanosized particles, and so on. A further consideration that should be noted in defining the precise scope of the ITAR is the existence of a "designation authority" afforded to the Department of State. This authority provides the State Department with the power to add to the articles and services that are restricted, essentially as it sees fit and with little direct oversight. That discretion may be exercised at any time with respect to almost any article or service that is intended primarily for military applications and that has "significant" military or intelligence capability.[70]

Because it is more directly concerned with military applications than is the EAR, the manner in which the restrictions of the ITAR are implemented are generally more limiting, even when the nominal scope of the restrictions is similar. For instance, the same kind of exception for exporting fundamental research that would otherwise be restricted by the EAR exists in the ITAR. The manner in which it is formulated is only slightly different, with fundamental research being one of several categories of information that is in the "public domain" under the ITAR. Other examples of information in the public domain include information available from such things as newsstands, bookstores, subscriptions, libraries, patents, and conferences.

Such public-domain information enjoys a general exception to the export restrictions imposed by the ITAR.

Despite this similarity in definition, the fact remains that the different regulations are put into effect by different government agencies that have different objectives and perceptions. The result is that the implementation of even these very similar rules is done in a visibly inconsistent manner. Even a single rule like that embodied in the ITAR is frequently implemented in different ways by the different regulatory agencies to which it applies. A good example that illustrates these different perspectives is made by Crocker,[71] who compares the mandate of the Defense Advanced Research Projects Agency (DARPA) to develop defense-related technologies with that of the Directorate of Defense Trade Controls (DDTC), which is charged with controlling the sales of arms to foreign powers. It is not surprising that interpretations of the ITAR restrictions by DARPA are generally more lenient toward exports than are the interpretations of DDTC.

The result of this pervasive inconsistency among different agencies puts nanotechnology researchers who wish to export technology in a difficult position. Compliance can really only be assured when those researchers abide by the most restrictive interpretation reached by any of the relevant agencies. Advice provided by one agency with a more liberal interpretation may not be sufficient to insulate against a finding by a more restrictive agency that an export was improper. This, in turn, results in an atmosphere that tends to inhibit exports, probably to a greater degree than the policy as promulgated by the government departments intends.

There is, of course, a third alternative to deciding not to export and accepting the risk that a different agency will find a particular export improper: seek a license. But this has its own difficulties. The procedure is time consuming and has no certain result. For example, if a license is required under the EAR, each of the Departments of State, Defense, and Energy must be consulted, as well as the Arms Control and Disarmament Agency. If an objection is raised by any of these, more investigation is needed that further protracts the process. Sometimes there may be disagreements between different departments, in which case an operating committee is convened to reach a final licensing decision. The process used when a license is sought under the ITAR is even more lengthy and bureaucratic.

Accepting the risk that an export will be found to be improper and proceeding with it is generally inadvisable. The penalties for making improper exports can be severe. Under the EAR, an individual who makes an improper export may be fined up to $250,000 and imprisoned for up to ten years, and an entity may be fined up to the greater of $1 million or five times the value of the exports. Under the ITAR, individuals and entities may be fined up to $1 million and individuals may additionally be imprisoned for up to ten years. These amounts are criminal penalties that may be imposed for each violation and may be supplemented with civil penalties.

### v. Office of Foreign Asset Control

Further restraints on exports can arise from a number of other government entities. One that was mentioned earlier is the OFAC, the Office of Foreign Assets Control, and some others are mentioned in one of the following Discussion points. The OFAC is an entity within the Treasury Department that is charged with the enforcement of economic and trade sanctions. As such, it has a more specific mandate than the departments that enforce the EAR or ITAR. While those two regulatory schemes are geared toward implementing a more general policy of maintaining control over information, the OFAC is engaged much more specifically in punitive practices against targeted countries or individuals.

Restrictions enforced by the OFAC are organized through the generation of detailed lists that identify—by name and alias—specific individuals or parties with whom all U.S. persons are prohibited from conducting transactions. Essentially, any U.S. citizen, U.S. permanent resident, U.S. company, or foreign national physically within the United States, is prohibited from having any dealings with the parties identified on the lists. They cannot give or receive funds, nor can they give or receive services, to (or for the benefit of) those parties. It is irrelevant what the size of the transaction might be, meaning that even the smallest of transactions may be prohibited. It is interesting that the penalties for violation of the OFAC regulations may be ever greater than the penalties for violating the EAR or ITAR regulations, reflecting the seriousness with which these sanctions are viewed. There is some variation in the actual range of penalties depending on which of several programs administered by the OFAC is violated, but criminal penalties can include fines between $50,000 and $10 million and imprisonment for between ten and thirty years for willful violations. Civil penalties currently range from $11,000 to $1 million for each violation.

Like other export restrictions, there are provisions within the OFAC for the granting of a license. While the administrative hurdles for obtaining a license are no less burdensome under the OFAC than under the other export regulations—and are almost certainly even more difficult to obtain—seeking a license to export technology with the OFAC is much less likely to be of importance to nanotechnology than under the other programs. The principal difference between these programs is the highly targeted nature of the OFAC arrangements. While it is easily conceivable that researchers might wish to communicate information about nanotechnology to foreign nationals in a manner that runs afoul of the EAR or ITAR restrictions, it is far less likely that there will be legitimate reason to do so with a party particularly identified by the OFAC. In academic settings, for instance, the EAR and ITAR restrictions impact the ability to collaborate fully with foreign-national students or researchers, but the scope of the OFAC restrictions is far narrower.

Like many regulatory issues, the landscape for complying with export restrictions is complex. Whether some interaction can take place with a foreign national is dictated by an array of different regulations that are not always consistent with one another in their intent nor in their application by different agencies. Yet export issues are of particular relevance to nanotechnology because of the manner in which it is being developed. In the book on workforce issues by Louis Hornyak in the Perspectives in Nanotechnology series, serious lapses in the ability of the United States to provide the technical personnel to develop nanotechnology effectively are identified. It is going to be necessary for researchers in the United States to solicit assistance from outside the country and this will necessarily require U.S. researchers to become familiar with the scope of export restrictions and to accommodate them.

### Discussion

1. There is a fundamental incompatibility between the various restrictions on the export of information and the patent system, which seeks to induce the general disclosure of inventions to the public. This incompatibility is addressed by allowing the patent office to issue licenses to export information for the limited purpose of filing a patent application in a foreign country. If a patent application is filed outside the United States without such a license having first been obtained, not only may the applicant have run afoul of various export restrictions, he will be prohibited from obtaining a U.S. patent on the invention.[72] Every application that is filed with the patent office is automatically considered for a foreign-filing license, with various government agencies being given the opportunity to evaluate the application as appropriate. In addition, a separate application for a license can be made; this is useful when no U.S. application is going to be filed or when there is insufficient time for the more usual process to be completed.

   It is important to keep in mind that a foreign-filing license granted by the patent office is not as broad as a license granted by the Department of Commerce or the Department of State under the EAR or ITAR. While those licenses will generally cover the technology as a whole, the patent office license is limited to that information truly necessary for preparing a patent application—it does not cover related information about the technology. This focus on patent filings also means that deemed exports to foreign nationals within the United States have not been authorized.

   Evaluate the effectiveness of the patent office foreign-filing license structure in balancing restrictions on the export of technology with the objectives of the patent system. Are there any changes you would make to the structure? How would those changes better balance these competing objectives?

2. It is sometimes incorrectly stated that the Export Administration Regulations apply only to "dual-use" technology, that is, technology that has both military and commercial applications. These

statements seem to imply that export of the technology is being controlled because of its military application, with the regulation of purely commercial applications being an incidental side effect. In fact, there are many technologies that have no evident military application that are still subject to control. What justification can you provide for controlling the export of purely commercial technology? What factors do you think are relevant in deciding to control purely commercial technology in this way? Pick some examples of nano-technology that you think have no military application and apply your factors. Under the guidelines you devised, would you restrict exports of any of that technology?

3. There are a number of other agencies in addition to those discussed in detail in this chapter that have requirements that may impact the ability to export nanotechnology. Many of these requirements are related to restricting investment in United States corporations. Research the role played by the Committee on Foreign Investment (managed by the Treasury Department) and the National Industrial Security Program (managed by the Defense Department). How do regulations administered by these groups impact the dissemination of information related to nanotechnology?

4. The nature of the OFAC list is sweeping. It can be violated with such simple transactions as selling a cup of coffee to a person named on the list or giving a quarter to a beggar named on the list. It is plainly unrealistic to expect every U.S. person to be familiar with the names of every person on the list and to check the identity of every person with whom they interact to ensure compliance. And yet the penalties for violation are extremely severe. Given the unrealistic nature of the restrictions, why do you think they are nonetheless structured this way? Can you devise an alternative scheme that you think is more realistic and accomplishes the same objectives?

---

## C.  Political and Judicial Control over Agency Action

In Carmel, New York, it is illegal for men to go outside with mismatched-color pants and jacket. Coins cannot be placed in one's ears in Hawaii. Duck parades are prohibited in McDonald, Ohio. You can be fined for leaving your couch in your garage in Cape Coral, Florida. And in Norfolk, Virginia, it is unlawful to flip a coin in a restaurant to see who pays.

Everyone is aware that there are numerous laws on the books that seem to have no rhyme or reason to them. The examples above are examples drawn from various local ordinances, but regulatory agencies are equally capable of promulgating regulations that seem completely inappropriate or that are excessively intrusive into the lives of individuals or into the practices of businesses. Examples of regulations that are viewed as unreasonable are not as outlandish as these very local laws seem (although it is perhaps worth noting that the context in which such laws were passed is not often considered when ridiculing them and such context frequently provides at least some plausible rationale for the law).

But regulations are still frequently promulgated that many view as unreasonable—security regulations that amount to a wholesale ban on an entire phase of physical matter being brought onto airplanes, environmental regulations that seem to favor the welfare of obscure species over the economic livelihoods of workers, and, perhaps most relevant to the discussion here, restrictions on the development and sale of foods that use genetically modified organisms that seem based on ignorance of the actual scientific issues. The degree to which nanotechnology is subject to regulation may well depend on a number of issues related to the way it is perceived by the public. Regulatory agencies are in many ways susceptible to pressure that may be exerted from public demands for regulation.

Perhaps the most fundamental concern with agency action is that it operates in many ways free from the encumbrances of direct political accountability. As discussed in some detail, agencies have significant power that has been granted to them by legislatures, and yet they are not subject to direct control by the populace through elections. There are instead two broad categories of control that are exerted over agencies to limit the exercise of power they have been granted: control by the political entities from whom their power flows and control by the judiciary to ensure compliance by the agencies within their political mandates.

### i. Political Control by the Legislature

Although agencies operate free of the direct control of the political populace, all of their power is ultimately dependent on legislatures who are subject to such control. There are accordingly two fundamental ways in which political entities retain and exercise control over agencies: direct control through legislation and indirect control through their funding power. Congress always retains the ability to revoke, modify, or change the scope of power of an agency, and has often done so in response to political pressures. Some relatively extreme examples are considered in the first topic of the next Discussion section.

Other examples include pieces of legislation established by Congress to require that certain considerations be accounted for when agencies take action. These reflect a certain posture on the part of the legislature that will also affect the way in which regulatory actions are taken with respect to nanotechnology. For instance, one concern of Congress is the impact that regulatory actions have on small businesses. The Regulatory Flexibility Act[73] directs agencies to evaluate the effect of their regulations on small businesses and to take steps to minimize their effects. Similarly, a concern about the environmental impact of agency actions is addressed with the National Environmental Policy Act[74] by requiring federal agencies to consider such impacts as part of their regulatory role. The Unfunded Mandates Reform Act[75] addresses a concern about the costs of regulation by requiring that rules promulgated by agencies that will impose significant compliance costs on private-sector or government interests be accompanied by a detailed economic analysis.

The funding power of legislatures can be just as effective in limiting the actions of regulatory agencies as such direct legislative control. In the United States, agency budgets are defined through an appropriations process in which agencies submit budget requests that are reviewed by the executive branch and transmitted to legislative committees for consideration. Such committees engage in a series of hearings to define spending bills that are ultimately voted on by Congress. More usually, the result of such a process allows Congress to use a generalized form of parsimony to limit the actions of particular agencies. But it is also possible for the appropriations process to be relatively focused. One example occurred in 1990 when Congress used an appropriations bill to temporarily modify several environmental statutes to permit logging in specific forests despite the threat such action posed to spotted owls (*Strix occidentalis*) in the forests.

Both legislative control and fiscal control over agencies may be considered to be forms of formal legislative oversight in which Congress uses its direct powers over the agencies it created. There are also a variety of informal mechanisms that Congress may use to retain control over agency action. These informal mechanisms are essentially grounded in the authority of the legislature to investigate the manner in which regulatory programs are implemented and to expose corruption or ineffective administration of those programs. This oversight may take different forms, but usually requires that the agencies submit periodic or special reports detailing their activities to Congress. In addition, Congress has established certain permanent organizations, including the Office of Technology Assessment and the Congressional Research Service, that monitor the activities of agencies. Sometimes, this monitoring function can provoke the exercise of a more formal form of agency control.

## ii. Political Control by the Executive

Control over agencies is not only affected by the legislative branch, however. The executive branch of government has the power to make certain political appointments to agencies, providing the president with the ability to populate certain key positions within agencies with individuals having a certain mindset or viewpoint. The potential scope of this power should not be underestimated. Simply contrast how different the regulation of nanotechnology could be with different people placed in authority at the relevant regulatory agencies: a scientist who embraces technological innovation is likely to direct an agency in a very different direction than is someone who is ill informed—or even misinformed—about the actual risks and benefits of nanotechnology.

The executive branch has other tools it may use to affect agency action. Of particular note is the ability of the president to issue executive orders. This power has occasionally been used in the past to influence agency activity and could be used in the future to affect the regulation of nanotechnology. A further executive power is the ability to control litigation that affects certain agencies through the Department of Justice. This is a very subtle form of agency control, but its potential power is quite remarkable. The ability to exert control over agency actions by the executive in this way derives from the fact that most agencies have no authority to litigate on their own behalf; they must instead rely on representation from the Department of Justice. When the executive directs that department not to defend a particular agency policy, the agency is left in a circumstance where its attempt to implement a policy is impotent.

The nature of political controls is such that there is a certain ephemeral quality to the kinds of pressures that are exerted over agencies. The political process is one that embraces compromise, bargaining, and negotiation to reach a certain acceptability of decisions. The resulting flux in policy that results from such a process is natural and expected. Every election has the potential to change the makeup of the legislature so that it has different priorities and every new president is generally expected to replace key political agency appointments with individuals who have views more consistent with his or her own. This fluid sense in the manner in which policy is implemented occurs even without any change in the underlying statutory mandate of the agencies.

### iii. Judicial Control

The nature of judicial controls is markedly different. The judiciary is—at least ideally—a nonpolitical arm of government that is less influenced by the vagaries of public pressure. Instead, judicial processes are predicated on the premise that actions taken by agencies are reasoned and result from a combination of effective fact finding and the application of established principles to the facts. Judicial control over agencies thus seeks to ensure that agency action is consistent with the will of the political branches of government as expressed through statutes.

Judicial review of agency actions is almost always invoked when a party brings a lawsuit alleging that the agency has acted improperly in acting on the authority given to it by the legislature. In the area of nanotechnology, this might take the form of a lawsuit asserting that certain regulations are too restrictive and therefore beyond the mandate of the agency. Or they might take the form of a lawsuit alleging that the agency is failing to regulate sufficiently by being too lenient. In either case, courts will look objectively to compare the manner in which an agency has acted with the expected protocol. This protocol has three fundamental prongs: the agency must interpret the law as enacted by the legislature, the agency must obtain facts about the situation it addresses, and the agency must apply its discretion in promulgating rules to apply the interpreted law in accordance with those facts. All three prongs may provide a potential basis for judicial interference with the actions of an agency and may therefore act as a mechanism for judicial control over the agency.

The nature of government structure is such that these three prongs are not treated with the same level of deference, implementing a careful balance between giving effect to the will of the legislature to delegate responsibility to the agency and to ensure that the agency does not exceed the bounds of that delegation. The judiciary is conventionally the final arbiter on what the law *is*. It therefore gives considerably less deference to an agency's interpretation of the law than it does to other factual conclusions or indeed to the manner in which it exercises discretion, as long as that discretion is exercised in accordance with a proper statutory interpretation. This balance seems appropriate. Judges are relatively objective with respect to the legal issues and have considerable experience and competence in applying well-established principles of statutory interpretation. But they are rarely sufficiently expert in the technical aspects of the agency's mandate. It is highly unlikely that a judge will have sufficient background or training to take issue with any but the most outlandish of factual determinations an agency makes about nanotechnology and the effects it may have on human beings, animals, or other parts of the environment.

None of this is to say that courts give no deference at all to agency determinations of what the law is. The statutes that assign authority to agencies are frequently complex and are directed to a defined area in which the agency has expertise. The agency is often very well positioned to understand how the different provisions of the statute interrelate within that area of expertise and has found ways of navigating various provisions in specific factual circumstances. There is accordingly a line of cases[76] that defines how much deference is to be provided to agencies with respect to their legal interpretations. The answer in a particular instance hinges on whether "Congress has directly addressed the precise question at issue." If so, and if it has been done in an unambiguous manner, the court gives very little deference to any deviation from the expressed will of the legislature by the agency. But, if not, the question is whether the agency has devised a reasonable interpretation of the issue. In areas where there are ambiguities in a statute that gives authority to an agency, the presumption is that the legislature intended for the agency to fill in the gaps in a reasonable manner.

This has significant effects for the regulation of nanotechnology. At the moment, virtually no statute assigning authority to an agency provides specific unambiguous guidance to the agency on how to regulate nanotechnology. It is up to the agency to interpret how to apply statutory mandates developed for other types of technology in this context, applying its experience with other technology to fill a gap left by Congress. Because Congress has not directly spoken to the issue yet, a great deal of deference is provided to the agency's interpretations. As long as the interpretation is not manifestly unreasonable, the agency is very likely to prevail with its interpretation.

But this still remains as only the first of the three prongs. The second prong still requires that the agency obtain facts about the situation it addresses so that a court reviewing an agency's actions must determine whether its factual findings may be sustained. There are different standards that a court might apply depending on the type of regulation being promulgated, that is, does it result from a formal trial-type procedure or is it the result of an informal rulemaking process. Informal processes may sometimes be reviewed without any deference to the agency at all, although this is rather unusual, or might be reviewed under a standard that seeks to determine whether the agency acted arbitrarily and capriciously. In most instances, though, a court will apply what is identified as a *substantial evidence* test.[77]

This standard is to be applied whenever an agency has promulgated a decision after a formal on-the-record type of procedure. The objective of the court is not to try to determine whether the facts accepted by the agency were in some sense "correct." This is an important point in evaluating agency actions in many technological areas and will undoubtedly be of concern with the regulation of nanotechnology. Concerns sometimes exist that agencies are motivated by political pressures to regulate technologies in certain ways and are vulnerable to a temptation to be selective in their fact-finding to support a particular politically motivated action. Application of the substantial evidence test should guard against such improper fact-finding activities because it seeks to assess the reasonableness of the fact-finding procedures used by the agency. But genuine mistakes by the agency in collecting facts are not subject to correction in this way as long as the procedure for collecting facts was reasonable.

The final prong in evaluating agency actions involves ascertaining whether the agency abused the discretion it is afforded. It is not merely enough that the agency have correctly interpreted the relevant law and engaged in a reasonably objective collection of relevant facts; it must also act in a way that is not arbitrary or capricious.[78] The focus on this evaluation is the quality of reasoning used by the agency to develop the promulgated rule. At its heart, this evaluation seeks to ascertain whether there is evidence that the analysis performed by the agency was at least plausible. Courts have identified four specific circumstances in which discretion has been abused.

First, a discretionary decision by an agency administrator that is inconsistent with its own rules is unlawful.[79] Second, departure from agency precedents is presumptively an abuse of discretion, although it is possible to overcome the presumption by adequately explaining a reason to depart from precedent. But there is a significant recognition of the fact that industry conditions sometimes change so fast that blindly adhering to agency precedent may frustrate rather than promote the objectives of the agency. Third, it is an abuse of discretion for an agency to breach principles of law as they are embodied in relevant judicial opinions. And, fourth, it may be an abuse of discretion when a remedy applied by an agency is too severe, particularly when the remedy is applied retroactively.

The ability for the judicial and political systems to exert control over the actions taken by agencies provides an important restraint on their power. At the moment, it remains fair to say that there is little specific direction provided to agencies that will have regulatory impact on nanotechnology. It will be up to each of those agencies to attempt to fill in these gaps that are left and many will disagree with how these gaps are filled—some will be concerned that a new form of technology has not been regulated nearly adequately, while others will feel that the regulations that have been promulgated are largely a result of ignorance of the true safety and usefulness of the technology. Political and judicial control provides a mechanism for ensuring that the authority does not acquire an unchecked life of its own.

### Discussion

1. After the development of nuclear power in the 1940s, the Atomic Energy Commission was responsible for the promotion and regulation of nuclear power. But in the 1970s, these functions were separated by Congress, with responsibility for promotion of nuclear power being delegated to the Department of Energy and responsibility for regulation being delegated to the Nuclear Regulatory Commission. Why do you think such an action was taken? More recently, the immigration services and enforcement functions of the Immigration and Naturalization Service were separated in 2003, with immigration services now being handled by the U.S. Citizenship and Immigration Services and enforcement functions being handled by the Department of Homeland Security. What do you think motivated Congress in each case to separate these responsibilities? Is this a rational response to the concerns that existed? Were those concerns valid?

2. Now consider the regulation of nanotechnology in the United States. Currently, a variety of different agencies have responsibility for aspects of nanotechnology simply because of the diverse way the technology manifests itself. But would it be sensible to have a centralized agency responsible for regulation of nanotechnology as a whole? What benefits would there be? What drawbacks would there be? Can you identify issues that might cause a public perception of a conflict within the functions of the agency that would lead Congress subsequently to divide those functions? Can you suggest a regulatory structure that consolidates regulation of nanotechnology but which might avoid such an occurrence?

3. The power of political appointment to agencies is significant. But the executive role in such appointment is diluted by the ability of

Congress to set qualifications for certain offices. What qualifications do you think should be required of those who head agencies that have a significant role in the regulation of nanotechnology?

## Notes

1. *New York Post*. March 31, 1978.

2. *New York Post*. March 30, 1978.

3. *Jackson (Mississippi) Clarion Ledger*. April 1, 1978.

4. The philosophical underpinnings of government structures that separate legislative, judicial, and executive power can largely be traced to the 1748 thesis of Baron de Montesquieu, Charles-Louis de Secondat, *The Spirit of the Laws*, Thomas Nugent (trans.) (MacMillan, 1949).

5. *Amalgamated Meat Cutters v. Connally*, 337 F.Supp. 737 (DDC 1971).

6. 5 U.S.C. §533(b).

7. 5 U.S.C. §533(c).

8. See Davis, Kenneth C. 1978. *Administrative Law Treatise*, 2nd ed.: New York: Panel Publishers. 6.8.

9. Weiss, Rick. August 16, 2004. 'Data Quality' Law is Nemesis of Regulation. *Washington Post*, p. A01.

10. Pub. L. 106-544, H.R. 5658 (Treasury and General Government Appropriations Act of 2001), §515.

11. 67 Fed. Reg. 8460.

12. See generally Conrad, J. W. Jr. 2002. The Information Quality Act— Antiregulatory Costs of Mythic Proportions? *Kansas Journal of Law and Public Policy* 12, 521–557.

13. *Guidelines for Ensuring and Maximizing the Quality, Objectivity, Utility, and Integrity of Information Disseminated by Federal Agencies*, 67 Fed. Reg. 8452 (February 22, 2002).

14. 5 U.S.C. §533(a) and §533(b)(3)(B).

15. Oberdorster, Eva. 2004. Manufactured Nanomaterials (Fullerenes, C60) Induce Oxidative Stress in the Brain of Juvenile Largemouth Bass. *Environmental Health Perspectives* 112: 1058–1062.

16. An earlier use of the word *Frankenfood* can actually be traced, but with a decidedly different usage: "It's almost as if a Dr. Frankenfood has created a customer creature with superhuman demands: To eat appetizers for dinner, entrees for appetizers, ethnic foods for breakfast and breakfast any time of day," Ryan, Nancy Ross. March 11, 1992. Reinventing the mean. *Restaurants & Institutions*.

17. Lewis, Paul. June 16, 1992. Mutant Foods Create Risks We Can't Yet Guess. *New York Times.*

18. See, for example, chapter 1 in Steve Jones's *Darwin's Ghost: The Origin of Species Updated* (New York: Random House, 2000) for an accessible review of the historical role of human beings in genetic manipulation of species. The examples discussed in the main text are drawn from this chapter, which also provides other examples.

19. See Sadrieh, Nakissa, and Espandiari, Parvaneh. 2006. Nanotechnology and the FDA: What Are the Scientific and Regulatory Considerations for Products Containing Nanomaterials? *Nanotechnology Law & Business* 3: 339.

20. FDA Forms Internal Nanotechnology Task Force. August 9, 2006. FDA Press Release.

21. In fact, the use of the mold in medical applications has a long pedigree, with evidence of such uses dating to ancient Greece and China. Observations of the specific effects of *Penicillium* date to at least as early as 1871, when Joseph Lister found mold contamination in urine samples prevented the growth of bacteria, and 1874, when William Roberts noted the absence of bacteria in cultures of the *Penicillium* mold.

22. McCormick, Joseph B., and Fisher-Hoch, Susan. 1996. *Level 4: Virus Hunters of the CDC.* Atlanta: Turner Publishing, see p. 349.

23. Weiss, Rick. November 23, 2006. EPA to Regulate Nanoproducts Sold as Germ-Killing. *Washington Post*, p. A01.

24. U.S.C. § 1314.

25. 33 U.S.C. §1318.

26. *NRDC v. EPA*, 822 F.2d 104, 119 (D.C. Cir. 1987).

27. Nanotechnology Briefing Paper: Clean Water Act. 2006. American Bar Association.

28. Id.

29. ABA SEER CAA Nanotechnology Briefing Paper. 2006. American Bar Association.

30. Id.

31. 42 U.S.C. §7415.

32. The Adequacy of FIFRA to Regulate Nanotechnology-Based Pesticides. 2006. American Bar Association.

33. Id.

34. 15 U.S.C. §2602(2)(A). Certain exclusions from this definition may apply, such as where pesticides are excluded and regulated under the FIFRA and food and drugs are regulated in the United States by the Food and Drug Administration.

35. Technically, application of the TSCA depends on whether a nanomaterial is a new chemical substance or is an existing structure used in a new way. This is, in fact, an important distinction in determining which precise provisions are invoked and at least some suggestion has been made that only those provisions directed to new uses of existing structures confer regulatory authority to the EPA. Regulation of Nanoscale Materials under the Toxic Substances Control Act. 2006. American Bar Association. This distinction is disregarded in the main discussion.

36. 15 U.S.C. §2605(a).

37. 15 U.S.C. §2603(a).

38. 15 U.S.C. §2606.

39. 40 C.F.R. §§ 720.3(cc) and (33), 720.36.

40. 40 C.F.R. §261.21.

41. 40 C.F.R. §261.22.

42. 40 C.F.R. §261.23.

43. 40 C.F.R. §261.24.

44. 40 C.F.R. §261.31.

45. 40 C.F.R. §261.32.

46. 40 C.F.R. §261.33.

47. RCRA Regulation of Wastes from the Production, Use, and Disposal of Nanomaterials. 2006. American Bar Association.

48. 42 U.S.C. §9601(14).

49. 42 U.S.C. §9602(a).

50. 42 U.S.C. §9604(a)–(d).

51. 42 U.S.C. §9604(e).

52. 42 U.S.C. §9604(j).

53. 42 U.S.C. §§9611, 9612.

54. 42 U.S.C. §9604(k).

55. 40 C.F.R. §261.4(b)(1).

56. Serrill, Michael S. September 15, 1986. Moscow Takes a Hostage, *Time Magazine*.

57. It is perhaps worth noting explicitly that merely having an adequate clearance level is not by itself sufficient to obtain access to classified information. Even those who have the appropriate clearance level must still have a legitimate need to access the information.

58. Executive Order 13292.

59. 18 U.S.C. §§ 792–799.

60. The U.S. Espionage Act has an interesting checkered history. It was originally passed on June 15, 1917, just after the United States entered World War I, and was later extended in 1918 with the Sedition Act. These laws were blatantly used as a mechanism to squelch government dissent in a manner that has become infamous. Perhaps the most notorious application of the Espionage Act was the imprisonment of Eugene V. Debs for making a speech that "obstructed recruiting" efforts during the war. Debs had stood for election as president in 1900, 1904, 1908, and 1912 before his arrest, and ran again in 1920 from his prison cell. As an inmate, he collected some 915,000 votes, a number similar to the total he collected in 1912. Debs was perhaps the most prominent, but was by no means the only high-profile figure imprisoned under the Espionage Act for criticizing the government of the day. Literally hundreds of conscientious objectors to U.S. involvement in the war were also imprisoned. The Espionage Act was also used in 1917 and 1918 to interfere with the distribution of newspapers and periodicals critical of the government. This was achieved by using the position of postmaster general to refuse to allow the U.S. mail service for delivery of identified materials.

61. 18 U.S.C. §794.

62. 15 C.F.R. §§ 730–734. The Export Administrations Act expired on August 20, 1994. Its lapse resulted in declaration of a national emergency by the president under Executive Order 12924 (August 19, 1994), followed by annual presidential notices extending the provisions of the Export Administration Act. It lapsed again on August 20, 2001, with the president then continuing its provisions through Executive Order 13222 (August 17, 2001) using the International Emergency Economic Powers Act.

63. This flowchart is adapted from Supplement No. 1 to Part 732 of the Export Administration Regulations (April 24, 2006).

64. 15 C.F.R. §734.5.

65. 15 C.F.R. §734.3.

66. 15 C.F.R. §732.1(b).

67. 15 C.F.R. §734.8(a).

68. 15 C.F.R. §734.8(b).

69. A more thorough discussion of this topic is provided in Crocker, Thomas E. 2005. Export Control Pitfalls for Government-Academic Partnerships in Nanotechnology. *Nanotechnology Law & Business* 2: 62–71.

70. 22 C.F.R., §120.3.

71. Crocker, p. 68.

72. 35 U.S.C. §185.

73. 5 U.S.C. §§ 601–612.

74. 42 U.S.C. §§ 4321–61.

75. 2 U.S.C. §§ 1501–71.

76. *Chevron U.S.A., Inc. v. NRDC*, 467 U.S. 837 (1984) and its progeny.

77. 5 U.S.C. §706(2)(E).

78. 5 U.S.C. 706(2)(A). It is perhaps worth noting that the notion of arbitrariness may sometimes also arise in the context of fact-finding as discussed supra. Despite the fact that there are similarities to the inquiry, these remain as separate and distinct parts of the overall evaluation of agency action. In the case of informal rulemaking, an agency's fact-finding process might be evaluated for a lack of arbitrariness and, in addition, its discretion in promulgating the rule in light of the ascertained facts may also be evaluated for a lack of arbitrariness.

79. Interestingly, this may be true even when the decision is in accord with governing statutes. See *United States v. Nixon*, 418 U.S. 683 (1974).

# III

## Liability

### A. Introduction

"I am become death, the destroyer of worlds." This phrase is an English translation of a line from the *Bhagavad-Gita*, an ancient Hindu scripture that recounts the instruction of Prince Arjuna by Krishna, an avatar of the god Vishnu. The *Bhagavad-Gita* is believed to be about 2500 years old and opens with Arjuna, an accomplished archer, riding his chariot onto the field of battle. When confronted with the fact that the ranks of the enemy include his friends, teachers, and relatives, he initially refuses to participate in the war. But ultimately, he is persuaded by the arguments and philosophy of his charioteer Krishna and joins the battle.[1]

The internal struggle faced by Arjuna is one that earned the empathy of J. Robert Oppenheimer (Figure III.1), the scientific director of the Manhattan Project in which the world's first nuclear weapons were produced. He reflected on the quoted phrase at the time of Trinity, the detonation of the first atomic bomb in the desert of New Mexico on July 16, 1945 (Figure III.2).

Since that time, the phrase has become forever associated with the destructive potential of science. While there are many examples throughout history in which the consequences of science have been negative, images of the devastation produced by atomic weapons in Japan at the end of World War II are undoubtedly among the most vivid.

### 1. Should Scientists Be Held Responsible for the Use of Their Creations?

There is much evidence that Oppenheimer's internal struggle, unlike that of Arjuna, remained unresolved until the end of his life. In the years after the conclusion of World War II, he made a number of comments that appear to attempt a justification for the role that he played in the destruction at Hiroshima and Nagasaki. The thrust of his argument is that development of nuclear weapons and their use was something in which many people played a role, all of whom had different strengths and areas of expertise. His own contribution was a scientific one, in which he did the best that he could in

**FIGURE III.1**
A photograph of J. Robert Oppenheimer taken late in his life, at the CERN physics laboratory
on the inauguration of its proton synchrotron in February 1960. (Credit: CERN, courtesy of AIP
Emilio Segrè visual archives. Image reproduced with permission.)

**FIGURE III.2**
Detonation of the world's first atomic bomb near Alamogordo, New Mexico, in July 1945.

developing the weapons that others deemed necessary to address the threat being presented.

It was these others who had the knowledge and training to determine how best to confront a threat by deciding what level of force was appropriate. It was not Oppenheimer himself who made the decision how to use nuclear weapons—there were, after all, a number of alternatives to direct attacks in Japan, including providing demonstrations of the weapons and seeking other ways to use them as a deterrent. Indeed, Oppenheimer might have viewed it presumptuous for him to have attempted to impose his own view on how they should be used because he lacked the necessary expertise—political, sociological, tactical, and so forth—to make such a decision.

This type of argument is one that is compelling to many scientists. It presents the scientist in a role in which he is absolved from moral responsibility for how his discoveries are used. Instead, this view argues that different duties are assigned to all those who were involved in the creation and use of technology. It is the duty of the scientist to confront the technical issues and "to turn over to mankind at large the greatest possible power to control the world;"[2] and it is the duty of the statesman to confront the societal issues in deciding how to use the technology. Oppenheimer aptly summed up his view by explaining that "Scientists are not delinquents. Our work has changed the conditions in which men live, but the use made of these changes is the problem of governments, not of scientists."[3]

This is also an argument that many find to be nothing more than a cowardly rationalization. In this contrary view, scientists are identified as the ones best positioned to understand the potential consequences of their discoveries, and they have an innate responsibility to consider the societal consequences of those discoveries. An attempt to defer a proper examination of those consequences is viewed as an indulgent excuse to explore interesting scientific issues without the constraints of ethical accountability. Indeed, other comments attributable to Oppenheimer also acutely reflect the existence of this attitude among scientists: "When you see something that is technically sweet, you go ahead and do it and you argue about what to do about it only after you have had your technical success. That is the way it was with the atomic bomb."[4]

These same kinds of ethical debates apply to nanotechnology. There is no doubt that as this technology continues to evolve, there will be a variety of instances in which it will have negative consequences. People will suffer injuries—and probably even death—that result from the use of nanotechnology products. In some cases, these injuries will be entirely accidental, with the producers of the products having had no way to anticipate that they would occur. In other cases, economic considerations may have caused companies to release products even knowing that there were certain risks associated with their use. In the most extreme cases, nanotechnology may actively be used to cause harm to others deliberately, for whatever motivations have prompted people throughout history to seek to harm others.

## 2.  The Application of Civil and Criminal Laws

The question that arises in these various circumstances is how to apportion responsibility for the harms that occur. Despite the fact that nanotechnology is an emergent technology, it reflects a lengthy pedigree that can help in answering this question. There are few forms of technology that have not had the capacity to be used improperly or destructively. And the existence of such a pedigree is manifested by a body of law that addresses how liability for injuries to others is to be allocated.

The next several sections consider how this body of law applies to nanotechnology under these different kinds of situations. Liability for harms may arise in both civil law and criminal law contexts. The fundamental distinction between civil and criminal law is that civil actions are brought by private parties—an individual, a corporation, or some other organized entity—while criminal actions can only be brought by the state. There are a number of practical ways in which this basic difference manifests itself.

For example, the nature and goals of penalties that may be imposed in civil and criminal cases reflect the differences in objectives of private parties and the government. Because private parties are primarily seeking to recover compensation for the injuries they suffer, the remedies available in civil actions are fashioned mainly to compensate the injured party for the loss. While it is possible for punitive damages also to be imposed, these are somewhat unusual in civil actions and are a departure from the main objective of civil actions to provide compensation.

This is in significant contrast from the penalties imposed in criminal actions, which reflect the objective of the state to control the behavior of the populace. In criminal actions, the penalties that are imposed seek primarily to deter others from behaving in ways that are deemed harmful to society as a whole. The penalties can be much more severe in criminal cases than in civil cases. While the imposition of fines in criminal cases are superficially similar to damages in civil cases in that they both require the payment of money, the payment in criminal actions is intended to be a form of punishment. Since it is not limited to compensating the injured party, it is frequently a much larger amount that would be imposed in a civil action. Also, the severity of punishments in criminal actions is not limited to financial punishments; the state may also deprive a guilty party of her liberty by imprisoning her and in the most severe cases may deprive her of her life by imposing a capital punishment.

The potential severity of criminal punishments is reflected in different standards of proof when compared with the standards in civil actions. In a civil case, a complaining party need only prove his allegations by a "preponderance of the evidence," that is, the allegation is more likely to be true than to be false. The proof presented need only tip the scale in the direction of the complaining party by the slightest amount for him to recover. In criminal actions, the allegations must instead be proved "beyond a reasonable doubt" or "to a moral certainty." It is not enough to show that the allegations are likely to be true; instead, sufficient evidence must be provided that the truth of the allegations is all but certain. There are, moreover, a number of individual rights that constrain the power of governments to prosecute individuals in criminal cases that do not apply to civil cases.

## 3. The Importance of Context

The basic principles for assigning and apportioning liability for harms are well defined. In many instances, it is possible for more than a single legal doctrine to apply and it is not at all uncommon for remedies to be sought under both civil and criminal law theories, nor even for multiple theories within civil or criminal law to be pursued. But in considering the discussion of how those principles apply to nanotechnology presented in the next several sections, it is worth reflecting on the overall perspective that context provides. In particular, it is worth deliberating on the general role of technology in society and whether the benefits that technology has to offer are effectively promoted with the compromise the law has settled on.

The role of context in considering these issues can, at times, be profound. If nuclear weapons had been developed and used to kill hundreds of thousands of people outside of a war and without the sanction of a political state, they would rightly be condemned as the actions of terrorist madmen intent on achieving a genocidal massacre. And the person who organized and led them to that action would be viewed as an extremist fanatic among the worst in history. But history instead provides us with a more balanced view of Oppenheimer, who nonetheless persists as an enduring symbol of the twin ability of technology to provide negative as well as positive contributions to society.[5]

Of course, the context in which the actions of Oppenheimer and the other Manhattan Project scientists must be viewed is an extreme one. But even if the scope of nanotechnology turns out to be more modest, it still provides a similar potential to impart both benefits and harms to society. Are those who provide technological solutions to society's problems appropriately held accountable when harms result? Are they instead given too free a hand to unleash insufficiently tested technology on a populace and punished too little when things go awry? Or are they instead hampered in their efforts to provide benefits to society by punishments so excessively severe that they inhibit scientists from contributing their most revolutionary ideas? Perhaps the most important question of all is whether the law provides adequate flexibility to recognize the different contextual circumstances within which harms occur and to permit the remedies for those harms to account for those different contexts.

### *Discussion*

1. Do you find Oppenheimer's justification for his contribution to the Manhattan Project compelling? If not, what mechanism would you suggest to compel scientists to conduct their research in a more

ethically sensitive way? Oppenheimer and the other scientists involved with the Manhattan Project were not only acting with political sanction but with an imposed sense of duty to contribute to the war effort. Would a political decision-making system that gave scientists more responsibility for social decisions involving technology be a step in the right direction or in the wrong direction?

If you agree with Oppenheimer, is there any scientific research that is ethically objectionable if it enjoys political sanction? What factors do you apply in discriminating between ethically acceptable and ethically unacceptable research under such conditions? How would you then prevent scientists from engaging in unethical research when the results of that research are demanded by the political state? Research the consequences Oppenheimer suffered when he adopted a position in ethical opposition to the state's desire to develop thermonuclear weapons (the "hydrogen bomb").

You may wish to consult the book in the Perspectives in Nanotechnology series on the ethics of nanotechnology by Deb Bennett-Woods in considering a framework to help answer these questions.

2. The text discusses actions that may be brought between private parties and by the government against a party in both civil and criminal contexts. What about actions brought against the government, such as to hold it responsible when it uses or supports nanotechnology in a way that causes harm? The doctrine of *"sovereign immunity"* prohibits lawsuits against a government unless the government consents to being sued. For almost all kinds of cases in the United States, both the federal and various state governments have consented to being sued, effectively leaving those areas where they cannot be sued as exceptions. Governments in the United States have mostly waived their sovereign immunity in the areas where lawsuits involving nanotechnology might be brought. Why do you think governments permit themselves to be sued in this way? What societal objectives are achieved by permitting such lawsuits? What factors do you think are relevant in deciding to retain sovereign immunity to lawsuits in other areas?

3. References are sometimes seen to *"victims' rights"* in the application of criminal law. Do you think this is a relevant consideration? If so, how do you reconcile your view with the fact that compensation remedies are intended to be provided through civil litigation, with criminal litigation being limited to addressing broader societal impacts instead of impacts on individuals? If you do not think the rights of victims are relevant to criminal prosecutions, do you see any danger that not affording specific rights to victims will dilute the relevance of context in determining punishments for crimes?

## B. Civil Liability

### 1. Application of Negligence to Nanotechnology

The principal role of courts is to determine what the law is and to apply it to the specific facts of a given case. While there is no doubt that this is a correct statement, the ease with which it is stated is extraordinarily deceptive. Both determining what the law is and applying it to a set of facts can be surprisingly difficult. The existence of ambiguities is frequently seen as incongruous with the law, which is instead often viewed as something that is relatively well defined. After all, documents that are "legalistic" are those that go on at tedious length—they define every term, they recite hordes of synonyms for a single concept, and they seem to try to cover every possible contingency. Indeed, they give every appearance of being at pains to remove every doubt as to what is meant. But rather than being evidence of the definiteness of the law, the detail provided in such documents instead reflects the reluctance of attorneys to allow the law to fill in gaps that might otherwise exist.

The basic reason that the law is, at least at some level, inherently ambiguous is because the circumstances in which it must apply are forever changing. Over any period of time, there are changes in the attitudes of citizens, the kinds of relationships parties have, political structures, and an endless number of other ways. Of particular interest here is the fact that there are technological changes that affect the kinds of harms that might be caused, the ways in which those harms might be remedied, the ways they might be discovered, and so on. But for the law to be confronted with a new issue does not even require that there be these kinds of changes since the facts of every case are themselves unique. The ways in which they differ from the facts of other cases might be great or minor, but even subtle differences can change people's perceptions about what the result should be.

Every new law student quickly becomes familiar with this ambiguity. As she learns a little bit about how the law addresses certain issues, and relates some of these interesting facts to friends and family, the new law student is quickly confronted with "what if" questions: What if the child who committed the act was a little bit younger? What if the will said "John Smith" instead of "my uncle"? What if the money the man picked up was on the sidewalk outside the door instead of on the floor inside the door? And so on. Initially, the new law student tries to answer these questions with what she has learned, but she quickly realizes that the permutations of the facts are potentially endless. As she continues her studies, she gets better at abstracting the principles that should apply so that it gets easier to distill what is relevant. (And eventually, she graduates and becomes like every attorney who answers such questions with a brief pause and a response that is much more sophisticated than it appears: "Well, it depends.")

The same process that the law student experiences—seeing more and more cases where decisions have gone one way or another and trying to extract a general set of principles—reflects the overall history of the *"common law."* This form of jurisprudence is found in all countries whose laws developed from the traditions of England—it is followed in all the states of the United States except Louisiana and in all the provinces of Canada except Quebec, both of which have strong French traditions. The distinctive feature of the common law is that it is not mandated by any legislative body; instead, it builds up over time as courts try to follow precedents established by prior decisions of other courts. The effort to follow such precedents is made with the goal of developing a consistent way of addressing similar kinds of issues.

In attempting to discern what the law is when discussing the common law, it is therefore necessary to examine prior decisions that have been rendered in the general area of interest. This is done with the goal of synthesizing a generalized framework from the way decisions are reached under different specific factual circumstances. The resulting framework can then be applied to new factual circumstances in a way that maintains consistency with the overall approach. It is conventional to express the synthesized framework in terms of a series of elements, essentially a checklist of factors that must be satisfied in order for a proper allegation of wrongdoing to be made. In the case of civil law, this is referred to as a *"cause of action,"* while criminal law refers to the analogous construction as a *"crime."* If even a single one of the elements is missing from the facts of a given case, there is no cause of action (in the case of civil law) or no crime (in the case of criminal law).

While criminal law was historically also a form of common law, it has more recently been codified in the form of statutes passed by legislators. Even so, the same structure that is used to specify causes of action at common law is still used in specifying statutory crimes by setting forth a list of elements that must be satisfied. By having a statute that sets forth a definition of a crime in this way, some of the synthesis that is required in determining what the common law is can be avoided. To ascertain what the law is, reference is made initially to the statute. But, even so, there are frequently uncertainties in how to interpret some terms used in the statute or how the statute should apply to certain kinds of facts. These uncertainties are resolved in the same manner as is done at common law, namely by looking to precedents to distill a generalized framework that resolves these kinds of ambiguities and then permits application to other facts in a way that maintains a consistent treatment.

One of the most flexible common-law doctrines—probably the most flexible—is *"negligence."* Like many legal terms, this doctrine goes by a name that has various nuanced nonlegal connotations. But when used in a legal context, the term refers to a specific set of elements that are required to be established in making out a valid cause of action. Those nonlegal connotations are irrelevant to the legal application.

At the heart of the doctrine of negligence is the notion that people should behave reasonably. If someone behaves unreasonably and that unreasonable behavior results in harm to another, the person causing the harm should compensate the person who suffers it. While this basic principle is easy enough to state—and sounds eminently sensible—there are innumerable ways in which difficult questions can arise. Several of these are discussed in the next sections in terms of the specific elements that define the cause of action of "negligence."

Before delving into the details of those elements, it is worth noting that the flexibility of the negligence doctrine makes it one that can apply to almost any sphere of nanotechnology. It is hazardous even to attempt to list the areas in which it might apply because such a list will necessarily be incomplete. The ability to apply the negligence doctrine is limited only by the ability of people to identify ways in which others have behaved unreasonably—and there is little shortage of that! But within this broad space of potential application, one area that stands out as particularly susceptible is the area of medical diagnostics and treatments. The negligence doctrine is already widely applied to hold physicians accountable for the results of their treatments in the form of *"medical malpractice,"* which is, in many respects, little more than a specialized form of negligence.

Examples of medical applications that are expected to use nanotechnology include a variety of techniques aimed at improved drug delivery. The promise that is held out with this kind of nanotechnology research is the ability to deliver very specific doses of drugs to exactly the desired locations within a body. The obvious application for such delivery is in the treatment of cancer. Conventional chemotherapy techniques take the approach of globally poisoning a person with a drug that has greater toxicity on certain kinds of malignant cells than it does on other cells of the body. But the impact of such a drug on even healthy cells is significant, leading to a variety of side effects that are difficult for cancer patients to tolerate.

One proposal for targeted drug delivery using nanotechnology is to bind drug molecules to one of the many attachment points of nanotechnology structures like fullerene molecules ($C_{60}$, $C_{70}$, etc.). This permits the drug to be delivered to a patient with the fullerene molecules, which have a size that makes it very easy for them to be moved within the body, particularly across biological membranes. Other nanotech-based drug-delivery techniques make use of encapsulation techniques to confine the drug within a cavity in a nanoparticle matrix. Of particular interest are nanoparticle structures that are biodegradable so that the drugs can be delivered under controlled-release conditions and protected from degradation as they are delivered.

These are examples of nanotechnology treatment approaches that are actively being pursued. The current level of sophistication in using nanotechnology diagnostically is probably greater than it is as a form of treatment. There are a number of nanotech biosensor designs that combine a biological element with a physical element. The biological element responds to certain conditions in the body in a way that causes the physical element to emit a signal. These nanometer-sized structures thus have the ability to signal the presence of certain biological conditions in a body.

Another example of a nanotechnology diagnostic technique involves using the fluorescent properties of certain quantum dots in medical imaging. The electronic structure of semiconductor quantum dots is sufficiently well understood that they can, without much difficulty, be fabricated to fluoresce at specifically engineered wavelengths. The wavelength at which they emit light is approximately related to their size, with smaller quantum dots emitting light at relatively short wavelengths like violet or blue, and larger quantum dots emitting light at longer wavelengths like red. Different combinations of colors can thus be created to provide a unique color-coded tag, allowing easy identification of specific tags by observing the color code. Quantum dots tend to bind relatively easily to sites and fluoresce more brightly than conventional imaging tags, providing an improved technique for medical imaging.

Still other interesting medical applications of nanotechnology include techniques that have larger scale effects. For instance, some interesting tissue-engineering techniques are being developed using nanotechnology in which a matrix of stem cells is prepared and delivered to an area having damaged tissue. The stem cells differentiate into the desired tissue type and provide replacement tissue to repair the damage. Use of this technique when the tissue is bone is an effective way to repair fractures. Nanotechnology is also being investigated to produce artificial cells that can be organized into tissues or even organs. Such artificial organs have the potential to replace damaged organs, offering a potential treatment for such conditions as kidney failure, liver failure, or diabetes.[6]

While these techniques clearly have exciting potential for contributing to revolutionary advances in medicine, it is equally clear that there are many things that are unknown about how biological tissues may actually respond. It would be foolish to deny the possibility of unexpected harms resulting from the use of these and other nanotechnology techniques on actual patients. It is important to recognize that this potential for causing harm is not the issue in evaluating whether a physician commits negligence by using these techniques. The law of negligence does not demand infallibility on the part of physicians, or indeed of anyone.

Instead, the relevant issues are whether the physician has acted in an unreasonable way and whether the harm that was caused is a result of that unreasonable action. The details of how these issues manifest themselves in actual application can be better understood in the context of the specific elements of a cause of action of negligence: duty; breach of duty; causation; and damages. To establish that a person is guilty of negligence, whether in a medical or other setting, it is necessary that each of these four elements be proved by a preponderance of the evidence.

### i. *Duty and Breach of Duty*

Although they are usually treated as separate elements of a negligence action, it is convenient to discuss the concepts of *"duty"* and *"breach of duty"* together since it is the combination of them that addresses whether a party has behaved unreasonably. The duty element requires that it be established in some way that the person accused of negligence owed a duty of care to the person suffering the harm. And breach of the duty occurs when that person fails to exercise that duty of care. In the normal course of events, this duty is simply to behave reasonably under the circumstances. But there are a number of instances in which the duty of care is heightened, and these are of relevance to nanotechnology applications.

Those who are offering services in a context where they have specialized knowledge or experience are required to exercise the degree of care, skill, and diligence that others having that level of skill and experience commonly exercise under similar circumstances. This heightened duty is most often ascribed to "professionals," such as physicians, attorneys, and the like. For this reason, it clearly applies to those instances where physicians are using nanotechnology diagnostically or therapeutically.

But when a harm results from some aspect of nanotechnology, it may apply more generally to anyone who has some nanotech expertise. Thus, in settings where nanotechnology scientists or engineers are providing services to others—offering advice on the installation of nanotechnology products in industrial settings or overseeing the operation of nanotechnology products—they too are likely to be subject to the higher standard. They not only need to behave in a way that would be reasonable for the average person, they must exercise the same care, skill, and diligence that other scientists or engineers with their expertise commonly exhibit.

Adhering to this higher level of care is perhaps best illustrated with an example. Suppose that a company develops a technique for producing coated buckyballs used in storing hydrogen. Such a technique could find tremendous benefit in efforts to generate hydrogen-based fuel economies, which currently suffer from difficulties in finding effective ways of storing the hydrogen. The company provides engineers who oversee the installation of equipment to provide hydrogen storage for customers. The engineers install this equipment in a way that conforms with all normal standards for the installation of hydrogen storage equipment. They have acted in every way as a reasonable person would under the circumstances.

But suppose that a paper had been published six months before in a widely read journal asserting that the way the buckyballs react with the coating material causes the particles to have much greater toxicity to humans than was previously known. The paper explains that there are specialized procedures that can be implemented to neutralize the toxicity without interfering with the other functions of the coated buckyballs. In this situation, it does not matter that engineers had not seen the paper and were unaware of these specialized procedures. In this example, suppose it can be proved that, as a group, nanotechnology engineers were aware of this research and had adopted implementation of the additional procedures to neutralize the toxicity. The engineers who oversaw the installation without using these procedures have breached their duty.

This example also illustrates the objective nature of the duty-of-care standard. The fact that the engineers were genuinely unaware of this research is irrelevant. The fact that they are normally diligent about reading this particular journal but an error at the post office caused that particular issue not to be delivered is irrelevant. It does not matter that the delivery failure was noticed, a back issue requested, and that the back issue had also not yet been delivered. In a variety of ways, it does not matter how conclusively they can prove that they were faultless in being unaware of this research. All that is relevant is that it had become common for others of their skill level to use the additional procedures, and that they failed to do so. The reasonableness of their efforts to keep themselves well informed are of no consequence under this heightened duty standard.

This same kind of analysis applies to harms caused from nanotechnology in medical settings. The level of care that a patient is entitled to receive is that level of care that would be provided by a certain kind of hypothetical physician. This hypothetical physician does not suffer from any of the idiosyncrasies that actual physicians possess and does not suffer from the lapses of memory, misunderstandings, or misjudgments that affect real physicians. Instead, this hypothetical physician always exercises the level of skill and diligence that physicians commonly exercise. At the same time, this hypothetical physician is not omniscient. He is not aware of truly obscure results that might be relevant to a particular procedure. He is also not omnipotent. He is not possessed with a godly level of skill in implementing procedures.

In ascertaining whether a breach of duty has occurred to the patient, the actions of the physician who administered the nanotechnology therapy are compared with what this hypothetical physician would have done. To state it somewhat simplistically, if they differ, the physician has breached his duty. It is worth reiterating in this context that there is no breach merely because the result did not turn out as the physician or patient expected. Nor is there a breach merely because the physician did not apply the best technique known to medicine—as long as that best technique was not one that had become commonly used among physicians. Instead, a breach of duty occurs simply when the physician did not apply a level of care at least as high as that of the hypothetical reasonable physician.

## ii. Causation

The fact that a party has breached a duty is not by itself sufficient to find that that party has committed negligence. Another element of the cause of action of negligence is that the breach has resulted in a harm. The harm itself can be to a person—usually a physical harm but potentially an emotional harm—or can be to property. But there must be some proof that relates the breach of the duty to the harm that was caused.

Consider this in the context of the earlier example in which engineers installed coated-buckyball-based hydrogen storage equipment. The mere fact that they failed to implement the toxicity reduction process does not make them negligent if no one ever suffers any harm from their actions. One consequence of this is that it provides a window of opportunity for them to correct their breach. When they become aware of the toxicity issue—by finally receiving the missing journal issue, by learning of it at a conference, or by some other way—they have a chance to return to the installation site and make corrections. Leaving this avenue open to them reflects sound public policy. If they had instead become liable at the moment of the breach, they may have had less incentive to make appropriate modifications.

The requirement of demonstrating a relationship between the breach of duty and the occurrence of a harm—proving causation—is one that presents a number of interesting challenges. There are two somewhat separate considerations that must be addressed in establishing such causation. First, the breach must be the "cause in fact" of the harm. This is universally presented in the form of a "but for" test: "But for" the breach, would the harm have occurred? If the answer to this question is "yes," there is no causation and accordingly no negligence. If the harm would have occurred even without the breach, there is no negligence.

This may be illustrated with a simple example: Suppose that a nanotech chemotherapy-delivery technique is prescribed by a physician to treat liver cancer in a patient. The treatment uses an encapsulation technique in which the encapsulating material is to be dissolved by certain liver enzymes, resulting in delivery of the chemotherapy drug directly to the patient's liver and not affecting other tissues. But it turns out during treatment that the encapsulating material is partly ineffective at holding the drug and some of it is absorbed by the patient's pancreas where it results in pancreatic failure. In this scenario, the pancreatic failure is a direct result of the nanotech treatment. "But for" the use of the technique, the patient would not have suffered pancreatic failure. If the physician also breached a duty of care in prescribing the treatment, he may be liable for having committed negligence.

This example also permits illustration of another doctrine, which is sometimes referred to as the *"eggshell-skull rule."* This is a doctrine that is not exclusive to evaluating negligence, and applies in a number of different civil actions. It essentially states that those injured by harms must be taken as they are. The allusion to an "eggshell skull" refers to the fact that liability for fracturing a party's skull cannot be avoided by arguing that the party has a skull so unusually thin that it is particularly susceptible to fractures. In the case of the patient with pancreatic failure, it does not matter that the patient may have had a rare condition that made her pancreas abnormally vulnerable to the chemotherapy drug. The physician is still considered to have caused the harm and may be liable for negligence if the other elements are satisfied.

In some cases, the "but for" test gives a result that feels unfair. This usually happens when there are two (or more) breaches of duties so that it is impossible to assign causation to just one of them. A simple way in which the "but for" test can fail in this instance is when it is easy to prove that one of the breaches caused the harm, but impossible to discern which one. A rote

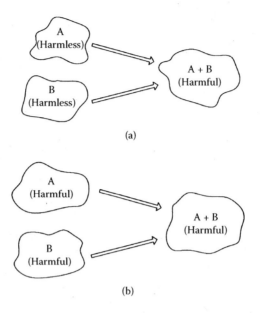

(a)

(b)

**FIGURE III.3**
Illustration of multiple actions giving rise to a harm. (a) Two harmless actions combine to cause harm. (b) Two actions each separately cause a particular harm. In either case, both A and B may be liable for negligence.

application of the "but for" test to each of the breaches independently results in a determination that neither one can be said with confidence to have been the cause of the harm. In other instances, the two breaches may have resulted in conditions that combined to cause the harm. Alone, either breach would have been harmless, so that it is impossible to say that the harm would not have resulted "but for" either of the breaches in isolation. In still other instances, both breaches are of the type that they may independently have caused the harm, again leading to a result where it is impossible to identify either of the breaches as "the" cause of the harm (Figure III.3).

In all these kinds of cases, what the law does is to impose a finding of causation on each of the parties who breached his duty. This allows the injured party to seek recovery from any or all of the breaching parties.[7]

In addition to requiring that the breach be the *"cause-in-fact"* of the harm, the law also imposes a requirement that the breach be the *"proximate cause"* of the harm. There are a number of different ways in which this concept is expressed, but all of them attempt to embrace the basic idea that the kind of harm that is caused by the breach should be reasonably foreseeable. Consider again the patient whose liver cancer treatment has resulted in pancreatic failure. Suppose that as a result of the pancreatic failure, the patient needs to be transported by ambulance to the local hospital. Because of this emergency transport, she does not show up to collect her son from school. He therefore walks home and is struck by a car while crossing the road.

Application of the "but for" test would find the physician who prescribed the nanotech chemotherapy treatment to be the cause in fact of the boy's injuries. If not for his prescription of this treatment, this whole sequence of events would not have unfolded and the son would not have been struck by a car. But it is clearly unfair to hold someone liable for a Rube Goldberg kind of sequence that results from his actions, even if those actions did constitute a breach of a duty. At some point, the harm becomes so far removed from breach that it seems wrong to impose liability.

How does one draw the line that demarcates the sphere of liability? One way is to begin by viewing every action as having an inherent risk. Then, the scope of that risk is defined by the kinds of injuries that might result from the action, the kinds of people who might be injured by it, and the ways in which those people might be injured. If one of these people suffers one of these injuries in one of these ways, the breach is the proximate cause of the harm; but if a different person suffers the injury, or if the injury is of a different type, or if the manner in which it is caused is different from the scope of this risk, the breach is not the proximate cause of the harm.

In the example being discussed, the risk associated with providing a medical treatment to someone is that there will be unforeseen side effects or complications that result from the treatment, and that they will affect the person being treated. Pancreatic failure in the patient is within the scope of this risk in every respect: the patient is a person within the scope of the risk; the patient suffered an injury of the type that is within the scope of the risk; and the patient has suffered that injury in a manner that is within the scope of the risk. This can be contrasted with the injury to the patient's son. That injury is outside the scope of the risk in every respect: the son is not a person reasonably at risk from the medical treatment to his mother; he has not suffered the kind of injury that could be foreseen from a nanotech medical treatment; and he has not suffered his injury in a manner that could reasonably be foreseen.

The requirement of establishing proximate cause provides a vivid illustration of the ambiguities that the law possesses, which were discussed at the beginning of this section. There is virtually no disagreement that the "but for" test is overinclusive in the sense that it would identify too many people as having caused certain injuries. But it is very difficult to define in a purely objective way exactly how to constrain that test. While tests have been proposed like the one described above to try to formalize a definition of proximate cause, the reality is that they are of very limited utility. In the majority of cases, it is clear as a matter of common sense whether there is proximate cause or not. In cases where the facts are close to the line, it is probable that a judge or jury often reaches a conclusion based on an abstract and ill-defined sense of what is right. They then try to explain that conclusion in the context of the test, rather than being guided objectively in the first instance by the test itself.

### iii.  Damages

The final element of the cause of action of negligence is damages. This element is directed at requiring a plaintiff to demonstrate some quantified measure of the harm that was suffered so that a court can award an appropriate sum of money. Because negligence is a civil action, the fundamental goal of fashioning a remedy for the harm is compensatory—the basic intent is to identify the financial value of any loss that the plaintiff suffered so that payment of that amount can be made by the defendant. It can be quite difficult in some circumstances to assign a monetary value to the harms that a person has suffered, particularly when those harms are physical.[8] Nanotechnology has the potential to harm individuals personally, such as when a nanotech-based medical treatment has an adverse physical effect, and to harm property. Assessing the financial value of damage to property is relatively straightforward—simply compare the value of the property without the harm to the value of the property with the harm and assign the difference as "damages."[9]

The situation is much less straightforward when a person has suffered physical harms. While even physical harms result in certain expenses that are fairly definite—medical expenses, lost wages, and so on—they also frequently involve financial impacts that must be estimated. A person who becomes disabled as a result of a negligently performed nanotechnology treatment may suffer physical and emotional pain, have a reduced future earning capacity, and need long-term medical or nursing care.

In assessing the financial value of these harms, what the law attempts to do is to put the person who suffered the harm in the same financial position she would have been in without having suffered the harm. This approach permits a reasonable assignment to be made of such things as lost future earnings by considering the different circumstances of the plaintiff before and after suffering the harm. A satisfactory estimate of a person's future earning potential may be made by considering the education level of the person, her age, her health, current employment, and so on. To the extent that these factors differ before and after the harm has been caused, they provide a good estimate of the financial loss that has resulted. Similarly, information is usually available to estimate what the cost of future medical and nursing care will cost, with expert physicians providing context by identifying the type and length of care likely to be needed.[10]

It is, frankly, impossible to assign any a financial value to the true intangible harms—loss of enjoyment of life, distress, grief, humiliation, and so on—using any formula, even in the abstract. There is simply no marketplace at which such things can acquire any definite value. Every person has different views of the world and different judgments on what things bring meaning to life. They consequently place different values on the loss of sight, the loss of use of a limb, and on every other physical and emotional aspect of their lives. It is true that different people also ascribe different values to tangible items, but the difference is that a marketplace exists for such items that permits an objective value to be assigned.

This fundamental fact is generally recognized in the law, which—for the most part—does not attempt to mandate an algorithm for assigning financial values to such harms. Instead, determining the value of such harms is subject to the discretion of a judge or jury, with the law imposing limits on the kinds of appeals that can be made to the judge or jury in influencing that discretion. The goal is to have such discretion exercised in a manner that properly and fairly accounts for the importance of these intangible harms, but in an objective manner. Overly emotional appeals are viewed as likely to inflate the awards.

This is perhaps most evident in the general prohibition on *"golden-rule"* arguments. It is impermissible in virtually every jurisdiction to ask jurors to imagine themselves in the circumstances of the injured party and to use that visualization in determining what award to make for noneconomic damages. Such an approach is viewed as too clearly being based on nonobjective factors.

Another type of argument that is problematic is the *"per diem"* argument. Such an argument asks jurors to assess the appropriate compensation for some small period of time—usually a day—and then to multiply that by the number of time periods the plaintiff will have to endure the noneconomic loss. While the amount for the small period of time may be modest, the total recovery may be considerable because of the long time that some plaintiffs will remain in their injured condition. Even though the per diem argument also tends to result in larger awards, it is not as universally rejected as the golden-rule argument. Indeed, in the United States, the majority of jurisdictions accept the per diem argument, with only thirteen states forbidding its use.

The existence of different levels of acceptance of per diem arguments shows that these arguments are at the periphery of the scope of discretion afforded to juries in assessing noneconomic damages. By any measure, that level of discretion is considerable. How that discretion is to be exercised in cases involving nanotechnology is not yet clear. Indeed, it is fair to say that there is still significant potential to shape public perceptions regarding nanotechnology. Because jurors are drawn from the general public, how those perceptions are fashioned will unquestionably influence the way noneconomic damages in nanotechnology cases are assessed. In the same way that Oppenheimer's position in history is framed by the context in which his actions are viewed, jurors will render their judgments according to the general context in which nanotechnology is viewed.

This is an important point that should not be lost on those in a position to influence public perceptions of nanotechnology. The effect of ensuring reasonable judgments in nanotechnology cases has the potential for widespread influence—a pattern of smaller damages awards results in smaller malpractice and other insurance premiums for those who use nanotechnology, which in turn permits them to offer their services at lower costs, and so on. This is just one reason why it is unfortunate that some popular discussions of nanotechnology have emphasized scenarios that are not only unrealistic but alarming. In light of what jurors currently know about nanotechnology, what is the result of a lawyer's argument along the lines of wondering "what would it be like to have to live out the remainder of your days knowing that you can never rid your body of the 'nanotechnology' that was put inside it and caused the plaintiff's condition"?

Thus far, this discussion of damages has focused on the kinds of factors that are considered in establishing the last element of a negligence cause of action and in determining what amount of money to award to a plaintiff. But a question that is at least equally interesting is "who pays it?" In the simplest cases where only a single party is the defendant, the answer is easy: it is the defendant (or, more usually, his insurance company) who pays. But there are a host of possible scenarios in which there is no single person who was negligent. The harm to a plaintiff may have resulted from breaches of a combination of duties by different parties. A simple example is where the manufacturer of a nanotechnology treatment product negligently produces it with some defect that a reasonable physician would have detected, but that was missed by the physician using it. In this scenario, both the manufacturer and the physician acted negligently in a manner that caused harm to the plaintiff. Other examples are those described earlier in which a simple application of the "but for" test does not cleanly correspond to causation and in which the law responds by holding multiple parties liable.

In circumstances where multiple parties are liable—described as *"joint tortfeasors"*—there are different approaches that are followed in different legal jurisdictions. The most traditional rule applies the theory of *"joint and several liability,"* in which each of the joint tortfeasors is liable for the entire damages. Essentially, the plaintiff is given the option of seeking recovery of the damages from any of the tortfeasors, without any assessment of their relative levels of responsibility for the harm. This is a remarkable rule. It is sometimes disparagingly referred to as the "deep pocket" rule because it has the effect of causing plaintiffs to add parties peripheral to the action based primarily on their ability to pay the damages.

In the scenario outlined above, suppose the manufacturer of the nanotechnology product is a well-financed corporation that has a value of $100 million and the physician is an independent practitioner who is covered by a medical malpractice policy having a $1 million limit and who has some $1 million in assets (the house he shares with his wife, his retirement accounts, some investments, etc.). In the negligence action brought by the plaintiff, the jury determines that the physician is 95 percent responsible for the injury and the nanotech manufacturer is 5 percent responsible, with total damages of $10 million. Recovery of this total amount from the physician is impossible, but under the traditional rule of joint and several liability, the plaintiff is entitled to recover the entire $10 million from the nanotech manufacturer.

That this kind of result is possible acts by itself as an incentive for plaintiffs to add marginal parties to the action in the hope that the jury will find some minimal contribution by those parties. The modern trend is to abolish or limit this doctrine. In some *"comparative negligence"* jurisdictions, the liability of each party is defined strictly according to that party's contribution to the harm—in the example above, the nanotech manufacturer would be liable for $500,000 and the physician would be liable for $9.5 million. In this example, this has the result that there is no way for the injured party to be fully compensated for the harms she suffered.

Many jurisdictions adopt an intermediate approach in which defendants found to have a level of fault below some threshold level have their liability limited. For instance, many jurisdictions impose a 50 percent threshold—if a party is less than 50 percent responsible, it can be held liable for only its pro rata share of the damages, but if it is more than 50 percent responsible, it may be held liable for the entire damages. Other jurisdictions have more complicated formulas, sometimes permitting double or four times the pro rata share of damages to be collected from defendants having a responsibility share less than some threshold value.

### iv. Defenses

There are a number of ways in which those who use nanotechnology may defend themselves against allegations of negligence. Typically, a complete defense to an allegation will take two approaches: it will attack the sufficiency of the case presented by the plaintiff and it will assert certain *"affirmative defenses."*

An attack on the plaintiff's case may take any of a variety of forms, ensuring that the plaintiff proves every one of the elements needed to establish a negligence cause of action. But within actions involving nanotechnology, there are two parts of the negligence action that may be particularly susceptible to attack: causation and breach of duty.

The vulnerabilities of the causation element arise from the fact that nanotechnology products, by their very nature, involve structures that are small and difficult to detect. The kinds of harms that are suffered from the use of nanotechnology products may well also result from a diverse assortment of other potential causes, including biological and chemical causes. In ensuring that a plaintiff has proved the causation element, it is well worthwhile for a defendant to identify other potential causes of the plaintiff's condition and to explore whether the plaintiff had been exposed to any of them. Particularly in the early stages of public encounters with nanotechnology, there will be a natural tendency for people to draw erroneous inferences that their use of a nanotechnology product is correlated with some condition that they develop. But the actual cause of that condition may well be much more prosaic.

Complicating this situation is the fact that tests for discriminating among the various potential causes are unlikely to be well developed, at least in the near future. Without an adequate test to decide with reasonable confidence that a particular condition is the result of any one of several causes, it may be impossible to differentiate among the different potential causes. In some instances, statistical techniques may provide relevant information to help

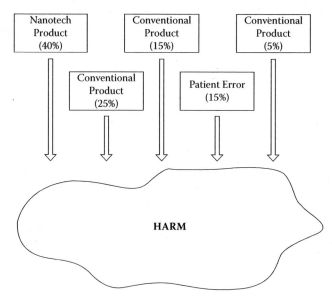

**FIGURE III.4**

A harm may arise from a number of distinct causes, with no single cause having a probability of causing the harm greater than 50 percent. Under such circumstances, how can the causation element of negligence be proven?

a plaintiff prove the causation element, such as when independent studies have shown that a certain fraction of a population that uses a particular nanotechnology product develops the condition at issue. If this fraction is relatively high in comparison with similarly derived statistical measures of other potential causes, it may be enough to meet the plaintiff's burden. Recall that the burden on the plaintiff is merely to prove the elements of negligence by a preponderance of the evidence—that the condition was more likely than not to have resulted from the nanotechnology product.

But if there are multiple potential alternatives, even if they individually are less likely to be the cause than the nanotechnology product, their collective effect is to reduce the overall probability that that product caused the condition (Figure III.4). Every additional legitimate alternative explanation that is offered, even those that represent remote possibilities, acts to reduce that probability. The net effect may be to raise a sufficient question about causation that a jury cannot properly conclude that the condition was more likely to have been caused by the nanotechnology product than not.

It is perhaps worth noting that attacking the causation element in this way is not just a form of attorney gamesmanship. Instead, this discussion illustrates an important facet of the proof requirements embodied in the preponderance-of-the-evidence standard. This standard is *not* met by showing that a particular cause is the most likely to have resulted in the condition. Rather, the standard requires showing that that particular cause *is more likely than not* to have resulted in the condition. In this way, the law reflects the statistical reality that there are circumstances (when there are multiple alternative explanations) where no one potential cause can be said to be more probable than the other potential causes combined. It is up to the plaintiff to demonstrate causation by offering proof that excludes alternative explanations so that the standard is met.

Over time, of course, better diagnostic techniques will be developed that will simplify excluding certain alternatives, making it easier for plaintiffs to establish causation. But the second attack on the elements of the negligence cause of action—that no duty has been breached—is likely to increase in viability over time, particularly in actions involving medical treatments. This element of a negligence action is of particular importance in medical contexts. The successes of medicine have unfortunately resulted in a somewhat pervasive view that if a medical treatment fails, it must have been administered improperly. Any sober reflection on the issue, though, recognizes that the failure of a treatment is very frequently nothing more than a manifestation of humanity's technological limits. Either we simply do not understand enough about the physiological response of individual human bodies to predict certain adverse reactions or the technology as it stands still includes inherent risks of failure or complications.

It is not enough to prove negligence by showing that administering a nanotechnology-based treatment resulted in someone being harmed, even when causation is proved conclusively. In medical contexts, having to prove the additional element that a duty has been breached usually amounts to a requirement that a physician must be shown to have departed from accepted medical standards. When a physician has used a nanotechnology-based treatment, such allegations can thus be countered with evidence that the physician, in fact, adhered to established medical protocols. This evidence might include the results of studies showing the general efficacy of the treatment, proof that regulatory bodies have approved its use for treating conditions like those of the plaintiff, journal articles recommending the treatment for such conditions, and so on. Of particular value in setting the context for determining whether a physician has breached his duty to a patient is the expert testimony of other physicians explaining why (or why not) the treatment was appropriate.

The nature of this type of evidence illustrates why the ability to attack this element of the negligence cause of action is likely to increase over time. As more and more nanotechnology treatments are administered, there will be better information available to physicians in deciding to use such treatments. There will be a greater number of studies evaluating the effectiveness and risks associated with nanotechnology treatments, and the studies will be more comprehensive. In short, the passage of time will see the creation and refinement of medical standards specific to the use of nanotechnology treatments. To be sure, there will always be occasions when physicians make errors or apply treatments outside of a standard in a fashion that amounts to the breach of a duty, but the existence of a reasonably well-defined standard will act to insulate those who comply with it from liability for negligence.

While these responses to the individual elements that the plaintiff must prove are frequently effective, the other class of defenses—*"affirmative" defenses*—take an entirely different approach. They basically accept that the defendant has committed negligence but effectively retort with a confrontational "So what?" They adopt a posture in which the negligence is acknowledged, but assert that there should nonetheless be no liability for it. There are principally two affirmative defenses to negligence: that the plaintiff contributed to the negligence or that the plaintiff assumed the risk of the negligence.

In actions that have a nanotechnology context, the more important of these is likely to be that there was some assumption of the risk. The underlying concept of this defense is that there was consent by the plaintiff to incur a known risk. Effectively, the voluntary and informed decision by the plaintiff to incur the risk acts to shift any duty of care to the plaintiff, thereby relieving the defendant of any liability for negligence. The two key considerations, then, are that the plaintiff knew of and understood the risk and that the plaintiff voluntarily accepted the risk.

When nanotechnology-based treatments are administered by physicians, this doctrine comes into play by ensuring that patients are fully informed about the nature of the procedure—specifically that it makes use of nanotechnology-based therapies—and of the level of risk that the procedure entails. An explanation of the risk should, at a minimum, include an identification of known adverse effects and the level of incidence of such effects. It should also include an explanation of the degree to which existing knowledge is incomplete. In the early phases of clinical uses of nanotechnology-based treatments, patients should be informed—to the extent that it remains true—that long-term studies of effects of the treatment have not been completed; that the studies that have been performed have been limited in scope, having been performed on demographically limited sets of patients or having been administered under protocols designed only to identify certain specific kinds of risks; that the physician himself is inexperienced in administering nanotechnology-based treatments; and so on. Usually some acknowledgment of having been informed in this way, and having voluntarily accepted the risks described, is signed by the patient and retained by the physician.

The failure of the patient to understand fully the nature of the risk may jeopardize a later ability of the physician to defend himself from an allegation of negligence by asserting that the patient had assumed the risk. Suppose, for example, that a seventy-two-year-old patient can establish that studies of a fullerene-encapsulation drug-delivery treatment had only been performed on patients under the age of fifty and that the treatment was later discovered to have harmful side effects in the elderly. The result of a legal action brought by the patient might well hinge on whether the patient was informed of the limitations in the studies, with the patient asserting that she never would have consented had she known that the treatment was untested in her age group. It is, of course, easier with hindsight to identify ways in which the information provided to the patient could have been more complete. But the more comprehensively that information is conveyed, the better will be the physician's ability to present a defense.

The other affirmative defense to negligence—"*contributory negligence*"—is based on the idea that a plaintiff should not to be entitled to compensation for another's negligence when her own actions contributed to the harm that she suffered. While this defense is likely to be less widely applied in actions involving nanotechnology, there are still a number of circumstances where it may arise, particularly in instances where a plaintiff has negligently failed to follow instructions for the proper use of nanotech products.

In many respects, issues involving contributory negligence can be considered to be special cases of joint-tortfeasor issues in which one of the tortfeasors is the plaintiff. The plaintiff owes a duty of care to herself and, if she has breached that duty in a way that is the actual and proximate cause of the harm she has suffered, she shares at least some liability for the damage caused by the harm with the defendant. And just as different jurisdictions apportion the liability in joint-tortfeasor circumstances differently, they also apportion the liability differently when the plaintiff has been contributorily negligent.

The traditional rule, which is still applicable in some jurisdictions, is that the plaintiff is barred from any recovery if contributory negligence can be proved. This is somewhat reminiscent of the joint and several liability assigned to joint tortfeasors, except that it is only the plaintiff who bears the entire liability for her own harm. Under this traditional rule, it does not matter what the relative levels of responsibility for the harm are among the different parties—even if the plaintiff is minimally responsible for contributory negligence, she cannot recover. The more modern trend is in line with the development of comparative negligence in apportioning liability according to a factual determination of relative levels of responsibility. In jurisdictions that have adopted comparative-negligence approaches, the liability of the defendant for his negligence is reduced to his level of responsibility. That is, a judgment that $100,000 in damages are 80 percent the result of the negligence of the defendant and 20 percent the result of contributory negligence by the plaintiff requires the defendant to pay $80,000 to the plaintiff. This can be contrasted with the traditional rule in which the plaintiff would recover nothing.

## v. Synopsis

This chapter, an "Application of Negligence to Nanotechnology," is one of the longest in this book. That length reflects the importance of the negligence doctrine and, in particular, the impact that it has on influencing the behavior of people in innumerable areas. It is not a form of government regulation that dictates specific forms of conduct that must be followed, but is instead a government-sanctioned mechanism that acts to promote reasonableness in the way that people interact. This is achieved by allowing the doctrine to function in a very symbiotic way with the scope of possible human behaviors: If a person behaves unreasonably and causes harm to another, the negligence doctrine permits that person to be held accountable for his behavior and to require him to provide compensation. In turn, others adapt their own behavior to avoid being subject to legal action, either because of a general apprehension about being sued or by noting specific results of lawsuits.

There are innumerable examples where highly publicized negligence actions have resulted in improved safety practices in entire industries. One of the most notorious is the case of *Liebeck v. McDonald's Corporation*, although the facts of that case have sometimes not been accurately reported. In February 1992, Stella Liebeck, a seventy-nine-year-old resident of Albuquerque, New Mexico, purchased a cup of coffee through the drive-through window of a fast-food restaurant. She was a passenger in a car driven by her grandson, who stopped the car so that she could add cream and sugar. She supported the cup between her knees while attempting to remove the plastic lid from the Styrofoam cup, but spilled the entire cup. She suffered third-degree burns in the area of her groin and upper legs and needed to receive skin grafts during an eight-day hospital stay. She wanted to settle her claim against the McDonald's Corporation for $20,000 to cover her medical bills, but her offer was refused with a counteroffer of $800.

Evidence at trial showed that the restaurant routinely maintained its coffee at a temperature of about 190°F, a temperature that causes third-degree burns in just seconds when in contact with skin and which its own quality-assurance manager testified was not fit for human consumption. Other evidence at trial established that coffee is usually served at about 140°F, that the restaurant company knew of more than seven hundred people who had previously been burned by its coffee over a ten-year period, including serious third-degree burns, and that numerous complaints by consumers and safety organizations about its coffee temperature had earlier been received by the company.

The jury found the defendant company negligent and also found the plaintiff contributorily negligent, apportioning their responsibility for the harm at 80 percent and 20 percent, respectively. The $200,000 damages award was accordingly reduced to $160,000. But the case's notoriety resulted from a jury award of two days worth of the company's coffee sales—$2.7 million—as punitive damages,[11] an amount that was reduced by the judge to triple the compensatory damages, or $480,000.

This case is not directly relevant to nanotechnology, except to illustrate the impact that a negligence action can have on safety practices. Later investigations showed that the fast-food restaurant had lowered its coffee temperature to 158°F, a temperature much more in line with prudent safety considerations.[12] This same kind of symbiosis should be expected of negligence actions that will be brought in actions that involve nanotechnology. People who suffer injuries as a result of nanotechnology-based treatments or products will hold those who have implemented that technology in an unreasonable way accountable. And it will cause those industries to reevaluate their practices, avoiding actions or habits that are likely to result in liability and implementing procedures that will mitigate the risk of causing harms.

In many ways, the potential for incurring financial liability in a civil context like this will be more effective at ensuring that safe practices are followed than will the dictates of regulatory bodies.

### Discussion

1. A criticism that judges often face is that they are "making" law rather than "interpreting" law. Assuming that there is general agreement that the former is impermissible while the latter is an indispensable judicial duty, what factors do you think distinguish between these two activities? When presented with facts that are clearly outside the specific contemplation of the express language of a law, what approach should a judge take to adjudicate properly? Can you suggest any effective oversight mechanisms to ensure that judges properly limit their decision-making processes to avoid "legislating from the bench"?

2. *Equity.* What is currently referred to colloquially as "the law" in the United States and other countries that follow an English tradition is actually a combination of two somewhat distinct kinds of legal systems that can trace their origins to practices implemented by English "courts of law" and "courts of chancery." Historically, legal remedies were sought from courts of law in England, which eventually developed a rigidity in which the types of claims they would hear were small in number and very restricted in their application. Many plaintiffs who had clearly suffered injuries at the hands of others were unable to make out a proper cause of action "at law."

   They accordingly petitioned directly for relief to the crown, which would exercise its residual judicial power to fashion an "equitable" remedy. As the law courts became increasingly rigid, the number of such petitions increased. The Crown at first delegated the responsibility to resolve the petitions to a chancellor. But as the number of petitions for nonlegal remedies mounted, a more formalized system of "courts of chancery" developed that would issue orders as a form of equitable remedy. The way in which remedies were fashioned by these courts developed as a judicial system in parallel to the "legal" system.

   There were a number of ways in which legal decisions and equitable decisions differed. And to this day, the distinction between *"legal"* and *"equitable"* remedies is maintained. Perhaps the most fundamental difference is that the only legal remedy that could be given was payment of money. Remedies that involved orders that a party perform or (more usually) refrain from performing some action—*"injunctions"*—were only granted by the equity courts. Other differences that persist include the basis for deciding cases, with legal actions being decided by rules that are generally more fixed than various "maxims of equity," which instead continue to embrace more clear notions of fairness: "One who seeks equity must do equity"; "Equity regards substance rather than form"; "Equity delights to do justice and not by halves"; and so on.

Construct examples of harms that might be caused with nanotechnology where you think monetary damages provide an inadequate remedy. Are there any generalities in your examples that you think illustrate more general principles where an equitable remedy is more effective than a legal remedy?

3. This section remarks on the basic role of civil remedies to compensate those who suffer harms. But it is possible for those whose harm is merely technical, that is, it has no readily quantifiable value, to recover *"nominal damages"* of $1 or £2. What societal role does the availability of such damage awards serve?

4. It is also possible under circumstances where the negligence is somehow aggravated—the defendant committed what is sometimes called *"gross negligence"*—for a plaintiff to recover punitive damages. The objective of such punitive damages is deterrence, in which it is hoped that increased damage awards will deter others from engaging in highly negligent behavior. But consider two identical plaintiffs who suffer identical injuries, one at the hands of a defendant who is merely "simply" negligent and the other at the hands of a defendant who is "grossly" negligent. The second plaintiff may be entitled to a considerably higher recovery than the first. What arguments can you present that this is a fair result? What arguments can you present that this is detrimental to the societal goals the availability of civil damages hopes to achieve? Which of your arguments do you find more persuasive?

5. Notwithstanding the efforts to ensure objectivity in awarding noneconomic damages to plaintiffs, some awards still sometimes seem to be plainly excessive. One proposal to address this is to impose a capped amount on the size of awards for noneconomic damages. Irrespective of your own view, what are the best arguments you can devise that this is not a sensible solution? What weaknesses are there in your arguments? Can you suggest a modified solution that is consistent with your arguments but avoids the weaknesses you identified?

6. The text suggests that an effective defense to a negligence action involving nanotechnology may sometimes involve a denial that there was any breach of an acknowledged duty. Can you construct any realistic hypothetical scenarios involving nanotechnology where the existence of a duty to the plaintiff by the defendant is in dispute? Now analyze your scenario in terms of whether the injury suffered by the plaintiff was the proximate cause of an action by the

defendant. Do you see any relationship between the duty element of negligence and the proximate-cause element?

7. The examples of nanotechnology research discussed in this section are intended to refer, at least for the most part, to the diagnosis and treatment of medical conditions that are either already being used or are likely to be ready for use in the relatively near future. There are a large number of more speculative proposals for medical applications that are also frequently encountered in discussions of nanotechnology. Some analytical discussion of a realistic way in which nanotechnology will develop in this field is provided in the book in this series that addresses the future of nanotechnology, by Thomas J. Frey. Do you think the overall framework of negligence is sufficiently flexible to apportion liability with such futuristic uses of nanotechnology? If you think it is deficient, give an example where you think a conventional negligence analysis gives an inappropriate result.

## 2. Strict Liability for Nanotechnology Products

Consider the parable that asbestos offers to nanotechnology. In modern times, the durability and heat resistance of asbestos were exploited in a wide spectrum of industries. While the most notable of these have been in the construction and automotive industries, asbestos was also widely used in a number of home appliances such as toasters, hair dryers, and coffee pots. It is estimated that some 3000 consumer products contained asbestos, exposing millions of people to the fibers that are now well known to cause fatal diseases that include mesothelioma and lung cancer. The time period over which asbestos-related diseases develop is on the order of decades. While hindsight examinations have been able to identify the occurrence of asbestos-related disease in ancient times, the long period for developing disease symptoms permitted asbestos to be ubiquitously used for much of the twentieth century. It was only in about the 1970s that concerns about the health dangers of asbestos exposure began to be taken seriously.

As increasing numbers of people presented disease symptoms that could be traced to asbestos exposure, a correspondingly increasing number of lawsuits were filed. The result has become what the U.S. Supreme Court characterized as an "elephantine mass."[13] To date, more than 700,000 claimants have filed lawsuits against more than 8000 companies. More than 70 corporations have gone bankrupt as a result, with studies showing that some 60,000 jobs have been destroyed and an estimated $10 billion in investment lost. Many fear that this is still just the beginning. Recent years have seen the rate at which new actions are filed accelerating. As direct manufacturers of asbestos products are forced into bankruptcy by the onslaught of litigation against them, the increasing number of claimants direct their efforts against producers of products that incorporated asbestos. Some hypothesize that there may ultimately be as many as 3 million claimants with a total financial impact of some $300 billion.[14] Once touted as a "wonder fiber," asbestos is perhaps now best known for spawning the longest mass tort in history.

While this represents the fate of a mature technology, nanotechnology currently remains as an emergent technology. But even so, there are already hundreds of commercial products being sold to consumers around the world that contain nanoparticles. Just some of the many examples are the use of $TiO_2$ particles that provide protection against ultraviolet rays in sunscreens and cosmetics, the use of $SiO_2$ particles as antiscratch additives in paints, the use of nanoscale silver particles in textiles or as coatings on children's toys to provide antimicrobial protection, the incorporation of fullerenes in golf drivers, and the production of dental adhesives with silica nanofillers. Food products are also in the early stages of being treated with nanotechnology, with antimicrobial surfaces being applied to improve preservation of food and by treating at least chicken and fish with nanoparticle vaccines instead of antibiotics. And the commercialization of nanotechnology remains in its infancy. Over the next few years, the various benefits that nanotechnology can provide will result in a steady, and probably accelerating, increase in the number of products that incorporate them.

The absence of significant reports of injuries resulting from the use of the various products has so far been extremely encouraging. But there are generalized concerns that the full scope of potential risks from the incorporation of nanotechnology into consumer products is not yet fully appreciated. The fact that nanoparticles are so small raises nonspecific concerns that they will have generally high reactivity and will easily be able to penetrate barriers within the human body to reach sensitive areas. Exposure to nanoparticles may arise from at least inhalation, ingestion, and dermal-contact mechanisms. The specific exposure mechanism from different consumer products may vary, permitting different transmission paths in the human body to be accessed.

### i.  Risks Associated with Nanoparticles

More specific concerns about the potential for nanoparticles to cause harm arise from studies of the relative toxicity of particles as a function of their size. It is a general geometrical feature of objects that as their size decreases, the ratio of their surface area to their volume increases. When the size decreases to the nanoscale, that is, where the presence of individual molecules of a substance becomes relevant, there may be surface-chemistry effects that are not present at larger scales. It has long been known that this generally results in progressively greater toxicity as particle size decreases.[15] There is also evidence of nanoscale particles having the ability to interfere with protein folding and to find effective transport to different organs in the human body through caveolar openings in membranes, in addition to a variety of other potential biological impacts.[16]

Studies of the potential toxicity of nanoparticles are continuing and there is no doubt that a much more thorough understanding of the specific nature of toxicity risks and ways to avoid them will result from such studies.[17] Other books in the Perspectives in Nanotechnology series, such as the one on environmental effects by Jo Anne Shatkin, discuss in more detail various specific effects that nanoscale particles may have. What is relevant to the discussion here is that there is a body of literature that identifies legitimate health risks associated with the kinds of nanotech structures currently being incorporated into consumer products. It is very likely that as those structures become increasingly ubiquitous, some consumers will suffer some form of harm from them. And it is absolutely certain that allegations of such harms will be made.[18]

There are a number of different legal theories on which product-liability actions may be based. Although the discussion in the previous section tended to focus on the application of negligence in malpractice settings, negligence does remain an important doctrine in pursuing product-liability claims. The application of negligence to product-liability claims is identical to that described in the previous section, with the need to establish that a duty existed toward the consumer and was breached. In practice, there are many instances where a consumer may be harmed by a product and where it is very difficult or impossible to establish the breach of a duty.

### ii. The Concept of Strict Liability

The more modern trend is therefore to impose *"strict liability"* on those parties who make consumer products available to the public—manufacturers, retailers, distributors, and so on. The concept of strict liability essentially does away with the need to identify fault in a party in order for that party to be liable. Instead, a party may become liable simply because it introduced a defective product into the commercial marketplace. The focus when different strict-liability theories are applied is on the product itself and not on any behavior by the producer or seller. The basic determination to be made, rather than whether a party breached a duty, is thus simply whether the product is in some sense defective.

The relative importance of negligence and strict-liability theories is almost entirely a function of the number of jurisdictions that apply these different theories in product-liability cases. The pedigree of strict liability is now well established: the State of California was the first jurisdiction to adopt this theory more than fifty years ago.[19] Since then, virtually every jurisdiction in the United States—except, perhaps notably, the federal government—has adopted some form of strict-liability theory and the trend appears to be for the adoption of such theories in Europe and Asia as well.

While negligence and strict-liability doctrines differ in their focus on how some expectation has been violated—that a product is not free of defects or that someone has not behaved reasonably—they still share certain elements that must be established to make out a cause of action. In particular, the causation and damages elements of proving liability for negligence are applied in essentially the same way in strict-liability actions. The plaintiff must prove that the defect in a product was the actual and proximate cause of her injury and that she suffered damages as a result. The principles used in determining how to quantify those damages are also applied in essentially the same way under the two doctrines.

The differences between the doctrines in identifying how expectations have been violated takes three principal forms: design defects, manufacturing defects, and warning defects. To recover under a strict-liability theory, a plaintiff must demonstrate that a defendant sold a product that suffers from one of these defects and, by implication, that the defect was present at the time of the product's sale. Each of these defects may be characterized in some way by reference to the way in which a particular product was designed. A design defect arises when the design of a product is, by its very nature, defective. A manufacturing defect arises when particular products deviate from the product design—when irregularities or flaws arise during processes used in manufacturing the products. This may occur because of some production error, because machinery used in the production processes have become uncalibrated over time, and so on. A warning defect arises when the design of a product has some inherent potential to cause harm so that a warning advising customers of that potential is warranted. But if the warning is inadequate in providing that advice, the product may suffer from a warning defect.

### a.   Design Defects

Because all three forms of defect are centered on the design of a product, it is perhaps most sensible to begin with a discussion of design defects. At the same time, this is undoubtedly the most difficult of the three types of defects to define in an objective way. The other types of defects are fairly well restricted in scope. Whether there is a warning defect focuses objectively on whether there are hidden dangers in a product and whether purchasers have been adequately advised of those dangers. Similarly, determining whether there is a manufacturing defect examines whether a particular product deviates from its intended design in a way that might cause harm. Both of these determinations, while necessarily still subject to the exercise of some judgment—in evaluating the adequacy of a warning or assessing whether some error has occurred in fabricating a product according to specifications—are at least clear in what kind of judgment needs to be made.

The situation is not at all as clear, though, in trying to assess whether the design itself is defective. In designing a product, a number of considerations need to be balanced against one another: economic considerations that will affect the cost of the product and its ability to be sold, aesthetic considerations that influence the attractiveness of the product to consumers, safety considerations that dictate the overall risk of harm associated with use of the product, durability considerations that impact the life cycle of the product, and a host of other factors. All of these factors are as relevant to the design of products that incorporate nanotechnology as they are to other designs. In asserting that a design is defective, what is essentially being done is to substitute one judgment of how to balance these various factors in place of the judgment that the producer made.

Almost always, this substituted judgment takes the form of alleging that insufficient emphasis was placed on safety considerations compared with the other considerations. Is it possible to have designed a product to be safer? The answer to this question is all but universally yes—no matter what product is being discussed and no matter how safe that product already is. But what is the cost of improving the safety of the product further? Does it result in only a marginal reduction in the risk of harm while incurring a huge cost or so reducing the lifetime of the product that it becomes completely unacceptable to consumers? Or is it possible to drastically reduce the number of injuries that the design will cause with only the most modest impact on the other factors? Is there some level at which products are considered "safe enough" so that designers should be given a completely free hand with respect to the other factors? Is a product safely designed if the risk of harm to the vast majority of consumers is negligible but yet of very high to a small part of the population (who might have rare allergies or other unusual sensibilities)?

The fundamental questions of what the ideal weighing of design factors is and how far a deviation from that ideal warrants denouncing the design as defective are difficult to answer. At the same time, it is important for the producers and sellers of products to have some objective means of evaluating the degree to which they are exposing themselves to liability through the use of a particular design. This is rightly of concern to such producers since a judicial determination that a design is defective acts to condemn an entire product line; it may also greatly expand the scope of liability as others harmed by the product learn that they have a cause of action against the supplier of the product.

Providing a measure of consistency in evaluating the designs of products is complicated by the fact that the universe of products is so large. Some products are intended to amuse toddlers while others are intended to be used in slaughterhouses by meat producers. The versatility of nanotechnology is such that it is not relegated to some corner of this universe, but has the potential to be included in products sold for a diverse array of purposes to a mixture of different kinds of consumers.

*Identifying Design Defects.* To provide the desired consistency in judging the designs of different products, the law has developed at least two principal tests: the consumer-expectations test and the risk-utility test. As its name implies, the consumer-expectations test focuses simply on whether a particular design comports with the reasonable expectations a consumer would have for such a product. To provide an extreme comparison as an example, contrast the consumer expectations for a child's toy covered with an anti-microbial nanotechnology coating and military weaponry that incorporates nanotechnology. Both products involve the use of nanotechnology, but consumers have very different expectations as to safety levels. Purchasers of children's toys expect such products to meet the highest safety standards, that the children they purchase them for as gifts for holidays or other celebrations will not suffer from lifelong respiratory illnesses because of exposure to the coating. But the purchasers of military weapons fully expect those products to be lethal. A similar acceptance of different safety levels exists with nanotechnology-based medical treatments. While consumers do not expect products used in such treatments to be lethal, they do generally acknowledge that risks associated with medical treatments are likely to be higher than risks associated with children's toys.

One difficulty that has become more widely recognized with the consumer-expectations test is its application to "complex designs," a difficulty that is perhaps especially relevant to product designs that use nanotechnology. One of the hallmarks of the consumer-expectations test is that it does not resort to evidence from experts—it is based on the everyday experiences of users of products. The concern with complex designs is that typical consumers lack sufficient expertise to assess the injury risks associated with the products.

In such instances, the preferred test in evaluating the design is the risk-utility test.[20] This tends to be true even in the decreasing number of jurisdictions that still recognize the consumer-expectations test for design defects. The risk-utility test recognizes that the decision of which factors to emphasize in developing a product design is essentially an economic one. The test accordingly seeks to apply a cost-benefit analysis that weighs the relative effects of providing greater safety to a design against the negative impact that such an alternative would have on other factors that bear on the design. A design is then defective if the cost of avoiding particular risks is less than the safety benefit that would result from such avoidance. This test has progressively gained greater acceptance as addressing the factors that are of most relevance in evaluating designs, resulting in a general displacement of the consumer-expectations test.

One issue that is of particular interest in applying the risk-utility test is the role of alternative designs. In early articulations of the risk-utility test, the availability of a technologically practical alternative design was typically one of several factors to be weighed in comparing costs and benefits of the design at issue. It was, at least in principle, still possible for a plaintiff to prevail even if he was unable to identify an alternative design that could practically have been implemented. The more modern trend is to view the existence of alternative designs as more central to the risk-utility test so that a comparison of the actual design of a product may be made against alternative designs. This reflects what seems to be an eminently reasonable view that it is essentially impossible to assess the cost of avoiding certain risks and to assess the safety benefits of avoiding such risks without a specific example that permits a comparison to be made.

In most jurisdictions, the risk-utility test now takes the form of showing that there were foreseeable risks of harm associated with the design at issue that could have been reduced by using a reasonable alternative design and that the failure to use that alternative caused the product not to be reasonably safe, that is, its risk of injury outweighed its utility.[21] There is generally no particular standard of proof required for showing the existence of an alternative, with prototypes, expert testimony, or even lay showings being sufficient under appropriate circumstances.[22] What is notable about this formulation of the test is its implicit recognition that not every product must use the best design. It is perfectly acceptable to use an inferior design, provided that the use of that inferior design still meets the minimal requirement of being more useful than harmful.

In considering alternative designs that involve nanotechnology products, one particular focus is likely to be on products that combine a traditional structure with a nanotechnology-based enhancement. Examples of such products are fabrics that bond fluorine whiskers to cotton fibers to produce stain-resistant clothing and glass surfaces made with nanoscale texturing that results in hydrophobic behavior to keep windshields clear. For the sake of argument, suppose that socks manufactured with nanosized silver particles are marketed as being odor resistant. The products are wildly successful, but it is later discovered that problems with bonding the silver particles permits them to be absorbed through skin on people's feet. A complex biological mechanism that was previously unknown results in accumulation of the silver particles in the gall bladder, increasing the risk of gallstone formation.

An allegation that these socks have a defective design could quite easily identify a traditional cotton sock as an alternative design. It is easy to demonstrate that the risk of harm would have been mitigated without including the nanotech structure and evidence could probably be presented to show that it was at least generically known that there are biological risks associated with exposure to particles of that size. Demonstrating that the product was defective then hinges on whether the socks were unreasonably safe—was the risk of gallstone formation outweighed by the advantage of reducing sock odor? While the answer might well depend on how high the risk actually is, it is easily conceivable that the advantage would be found to be incommensurate with the risk.

This same type of analysis could easily play out with countless consumer products, with the benefits provided by including nanoscale structures being outweighed by harms that are discovered to result from exposure to those structures. Because it is so simple in such cases to identify a reasonable alternative design, proving the existence of a design defect is straightforward.

Importantly, there are certain products where the analysis is not nearly as easy. This is true in those areas where the use of nanotechnology is enabling the production of completely new types of products. A good example is the production of targeted drug-delivery products that make use of fullerene structures. If the fullerenes are found to cause certain harms as a side effect to the treatment, application of the risk-utility analysis does not lead as clearly to a finding that the design of the products was defective. The absence of viable alternative products for delivering drugs in such a targeted fashion is critically relevant in determining that the design is not defective, notwithstanding the fact that it has the potential for causing harm. This is true even in cases where the risk of harm might be significant. Nanotechnology products used in medical treatments are far more likely to result in this kind of analysis than are common consumer products. In some ways, this difference simply confirms the observation made earlier in the context of a consumer-expectations analysis—the different natures of these kinds of applications allows higher risks to be tolerated for medical applications than for consumer products.

*Defenses.* Some of the defenses discussed previously in connection with negligence may also be asserted in products-liability actions, including actions that allege the presence of a design defect. The most important of these asserts that any risk involved with using the product had been assumed by the plaintiff. The analysis of such an *assumption of the risk* defense is similar in this context to that described for negligence. A person who is adequately informed about the risks of using a product and willingly accepts them cannot later try to shift liability if they result in an injury.

The parallel defense of contributory negligence is generally unavailable in strict-liability actions, although jurisdictions that have adopted comparative-fault principles may reduce liability even for design defects when some allocation of fault among the parties is possible. A related defense that is more successfully raised in design-defect cases is that there was misuse of the product. While this bears some similarity to the idea that the plaintiff contributed to the injury, it is not necessary to show that there was any negligent breach of a duty that the plaintiff owed to herself, as there is in asserting contributory negligence. Importantly, the defense of consumer misuse is limited to *unforeseeable* misuses. This is because products are expected to be designed in a way that makes them reasonably safe not only for intentional uses but also for foreseeable misuses. And this limitation on the misuse defense extends to alterations or modifications that the consumer may have made to the product. If they are of a foreseeable nature, the fact that the product was misused (or altered or modified) is not enough to prevent a plaintiff from recovering.

Another defense that is worth noting when discussing nanotechnology is the *"state-of-the-art defense."* This defense essentially presumes that the safest products are those that incorporate the most advanced technology available at the time they are made. There is an evident logical fallacy in this rationale in that it is possible for advanced technology to make products less safe rather than safer. Indeed, in recent years, this defense has been weakened as a result of being interpreted in light of the need for the plaintiff to identify a reasonable alternative design. The state-of-the-art defense is now viewed primarily as little more than the common-sense recognition that the incorporation of cutting-edge technology into a product may make it difficult for a plaintiff to identify a reasonable alternative design. The result is a general rejection of the notion that a product's design is not defective merely because it makes use of state-of-the-art technology. For these reasons, this defense is, in fact, quite unlikely to find any special application in considering nanotechnology products.

The defense that is perhaps most likely to bear significantly on nanotech issues asserts that a certain product was *"unavoidably unsafe."* This is particularly the case when the nanotech product was designed for medical uses. The "unavoidably unsafe" doctrine has traditionally found its most effective application in medical contexts, again reflecting the special tolerance for risk that products-liability law has in medical settings. And there is every reason for it also to find widespread application in defenses related to nanotechnology. The defense rests on the recognition that there are certain products that have legitimate utility but are simply of a nature that causes them to be inherently unsafe. As the effects of certain nanotechnology-based medical treatments progress, there may be products that fall into this category, but still have sufficient utility that their use remains warranted. A frequently imposed prerequisite to enjoying this defense that should be recognized, though, is that the plaintiff was warned of the risks associated with the product. In this respect, this defense frequently has some overlap with the "assumption-of-the-risk" defense in that a user implicitly assumed the risk by proceeding in face of the warning.

### b. Manufacturing Defects

The issues surrounding manufacturing defects are much more straightforward than those related to design defects. This is because there is much less of a need to balance nonquantitative factors in determining whether a product is defective. The relevant inquiries in these cases are simply whether a specific item deviates from its intended design and whether that deviation resulted in the injury.[23] What is emphatically not relevant to the inquiry is the level of care that was exercised in producing the item. It is irrelevant that the product was produced in a model state-of-the-art factory with the highest safety standards and multiple levels of the most stringent quality-control checks ever implemented in any industry. Just as it is irrelevant that the product was produced in a facility where unqualified workers who barely understood the product specifications were forced to manufacture the products during excessively long shifts under intolerable conditions without supervision. The one-in-a-billion product that makes it through the first factory with a defect presents the same liability concerns (for the manufacturing defectiveness) as the thousands of products churned out every hour by the second facility with the same defect.

What is also important to recognize is that this absence of a fault-based analysis causes the potential for liability to extend even to parties who are absolutely certain not to be the ones responsible for introducing the defect. This circumstance may arise when multiple parties are involved with the manufacture of a product. For example, suppose a manufacturer of some sporting goods—say a tennis racket or a baseball bat—manufactures its items using materials that incorporate fullerenes to provide increased strength at a lighter weight. But a shipment of the material is defective in a way that is not detected by the sporting-goods manufacturer. When it sells its products, they turn out to be brittle, shattering when used and causing a variety of injuries as sharp debris impacts players. It is, of course, possible for an injured party to recover from the supplier of the nanotech material. But it is also possible for the injured party to recover from the sporting-goods manufacturer. It is even possible for the wholesaler who purchased the product from the manufacturer or the retailer who purchased the product from the wholesaler to be liable. Generally, any party in the manufacturing or selling chain who transferred a product containing the defect may be held liable.

### c.   Warning Defects

The discussion of products liability and the discussion of negligence both noted that there are circumstances in which warnings should be provided to alert consumers of potential risks associated with products. But not every warning is sufficient. Contrast the following warnings about the application of paint that contains nanoparticles: "caution" versus "warning": Apply only in well-ventilated area. Studies of the toxicity of nanoparticles in paint have been inconclusive" (Figure III.5). If the second warning is accurate, it is the better warning. The first warning is defective because it gives no information about what the risk is, how the risk can be mitigated, or what the potential consequences are.

One of the underlying rationales supporting the use of warnings is that there is some level at which the consumer should be the one to take responsibility for the risk that a product may pose. The interplay of this rationale with the idea of design defects is an interesting one. Another approach that society might have taken would have maintained liability for any unsafe product on its manufacturer. But that approach would result in many useful products not being released to consumers because the potential liability on manufacturers was too high and the product insufficiently valuable to justify extensive and costly safety redesigns. The ability to disclose risks to consumers, who may then exercise their own judgment in deciding to use a particular product, is an intermediate approach that has greater economic efficiency.

But if consumers are to accept the risk of using a product, they should do so from a position of reasonable knowledge. Manufacturers of products

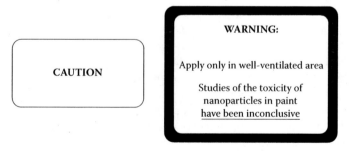

**FIGURE III.5**
A comparison of two warnings that might be affixed to paint that contains nanoparticles. Which is the more effective warning?

are expected to provide *adequate* warnings of *foreseeable* risks. This requirement again embodies the common theme in assigning liability that the law
does not require omniscience. The failure to provide a warning about risks
that were not only unknown to the manufacturer but not even reasonably
foreseeable will not result in liability. Whether a given warning meets the
adequacy requirement is judged according to whether the warning is clear,
specific, and communicates some understanding of the degree of risk. In the
paint example, very different warnings should be given if the manufacturer
knows that the only risk is of temporary lung irritation with no long-term
effects as compared with the manufacturer knowing that there is a risk of
long-term debilitation or even death in sensitive individuals. Warnings that
are vague in specifying what the risk is or are difficult to understand are
likely to be found inadequate.

The adequacy of the warning is also not judged merely by its content. It
is also important that the warning be conveyed in a way that is effective
in reaching users of the product. It should be displayed in a conspicuous
place and manner, using such features as color, type size, and eye-catching
symbols to enhance its visibility. Warnings that are located in positions
where they are unlikely to be noticed are inadequate even if the content of
the warning is ideal in content. Like an evaluation of a warning's content, the
prominence of a warning is also judged in the context of the level of risk. In
cases where the risk is higher or where it is more critical that the product be
used in a certain way, the more noticeable the warning should be.

Another factor to be considered in evaluating the adequacy of a warning
is the level of sophistication of the product user. Those who use common
consumer products in which nanotechnology structures might be implemented are not expected to be particularly knowledgeable about the potential toxicity of very small particles. When marketing products like golf balls
that incorporate nanotech structures or clothing that incorporates nanoscale
particles, the issues surrounding warnings should be addressed in the context of a relatively unsophisticated consumer. The same is not necessarily
true when the market for a product is generally more knowledgeable. For
instance, nanoparticle suspensions of zeolites prepared for research appli
cations might be marketed to scientists who are much more familiar with
the potential toxicity issues and accustomed to taking suitable precautions.
Warnings for such products can differ in every respect—by being less
detailed and less prominent—and still be considered adequate because of
the difference in user sophistication.

### iii.  Learned Intermediaries

The *learned intermediary* principle extends to circumstances in which distribution of a product to an end user customarily proceeds through a sophisticated intermediary. The learned intermediary doctrine finds particular
application in medical contexts in which a product is intended for use by a
patient, but only under the guidance of a trained physician. Medical devices

that incorporate nanotechnology will frequently fall under this categorization, with the producers intending them to be used by patients but engaging in commercial distribution to physicians.

A warning provided with products distributed in this way is adequate if it conveys the necessary information to the physicians, even if it would clearly be inadequate from the perspective of the average patient. The physician has her own duty when considering the use of such a product in treatment to be knowledgeable about its benefits and risks and to convey that information within the context of an overall treatment plan to the patient. In many instances, the failure of the patient to be sufficiently well informed of the risks associated with the product may give rise to a cause of action in negligence against the physician rather than a warning-defect action against the manufacturer.[24]

## iv. Conclusion

At the risk of sounding paternalistic, the discussion of warnings raises a further issue that the parable of asbestos presented at the beginning of the "Strict Liability for Nanotechnology Products" section also speaks to. A significant factor in the success of plaintiffs in asbestos litigation has been the ability to prove that producers of asbestos products actively suppressed information they possessed about the health dangers posed by their products. This is also true of a number of other high-profile products-liability actions, perhaps the most notable being recent actions against producers of tobacco products.

The reasons for suppression of risks are always the same. With "unavoidably unsafe" products perhaps illustrating an exception, the disclosure of dangers associated with products almost always results in reduced consumer acceptance of the product and demands that its safety be improved. The consequences this poses to the producer are increased production costs and reduced sales, both of which cause a reduction in profit. Sometimes producers will rationalize suppression of safety information by noting the existence of uncertainty in their evaluations.

But what the asbestos litigation demonstrates is that that analysis is incomplete. Consider how successful asbestos litigation would have been if there had been more frank disclosure of risks in the form of warnings presented to consumers. In such circumstances, the existence of uncertainty would have acted directly in favor of the producers. Because the actual risk level was uncertain, there would still have been a significant class of consumers who would have purchased and used asbestos products after performing their own balancing of risks and benefits. But such consumers would now be in a position of finding it difficult to make a products-liability cause of action under either a warning-defect or design-defect basis.

There would be no warning defect because the consumer was adequately notified of the known and foreseeable risks associated with the product. And, interestingly, it would be very difficult to recover under a design-defect theory. Even if an application of the consumer-expectations test or the risk-utility test forced a conclusion that a particular asbestos product was defectively designed, there would be a solid assumption-of-the-risk defense available to the producer. By informing customers of the risks and permitting them to make their own assessments of whether accepting those risks was warranted, much of the liability could be avoided.

A proper calculus of suppressing risk information thus not only accounts for the short-term impact on such things as production costs and market penetration. It also accounts for the long-term impact of liability costs. With a complete assessment of such costs, it may well be that the more economically prudent action is to disclose the risks. The history of asbestos litigation demonstrates that a failure to account for such costs may well lead to the demise of companies involved in the production of unsafe products. Certainly the objective of structuring products-liability law in the way it has developed is for those liability costs to be large enough that they act as an inducement for the behavior that society deems desirable: the disclosure of known risks and the promulgation of effective warnings to consumers.

Those involved in the development of nanotechnology products would be wise to learn from the lessons that asbestos litigation has to offer. The economic (and social) value of conveying warning information to the public if and when risks are identified is but one of these lessons. The scope and complexity of asbestos litigation are so great that they offer innumerable other forms of insight into how to conduct the prudent marketing of products for those willing to uncover them.

To express it differently, nanotechnology is still at a stage of development where the capacity remains to frame the context in which it is viewed. And, as with all technology, this context is intimately tied with the way responsibility for the effects of the technology will be viewed by others. If actions are taken to frame the context in the right way, it can mean the difference between fostering a sharply negative view that will result in a tendency to impose significant liability and a more positive view that will soften the potential for liability.

### Discussion

1. How does "strict liability" differ from "absolute liability"?

2. This section has characterized design defects in terms of strict liability. But how does a determination that a product is made with a defective design differ from a determination that the producer of the product behaved unreasonably in releasing that design to the public? Can liability for design defects be better characterized in terms of a negligence theory? Provide such a characterization.

3. In discussing the consumer-expectations test for design defects, it was noted that consumers not only accept that some nanotechnology products might be lethal, but expect that to be the case. Military weaponry is an example. But does this expectation allow producers of such products to completely ignore safety concerns? Does a

nanotechnology-based weapon that causes widespread environmental damage, injuring people well outside a defined target zone,
suffer from a design defect? What about a hand-held weapon that
kills the operator of the weapon about 5 percent of the time? How
do you explain your answers in terms of the consumer-expectation
test? In what ways does the risk-utility test provide a better analysis
of these kinds of circumstances?

4. The need to show the existence of a reasonable alternative design as
part of establishing a design defect has been criticized as being an
unreasonable demand imposed on certain classes of plaintiffs. Do
you agree? If so, construct an example involving nanotechnology that
illustrates how the requirement becomes so excessively burdensome
that it might prevent a legitimate recovery by an injured party.

5. A defense to a design-defect action that was not discussed is that the
defect was *"open and obvious."* The basic principle behind this defense
is that sometimes a defect can be so plainly apparent that the harm
to the plaintiff is viewed as being his own fault for not using adequate caution. The trend is for this doctrine to become weakened,
being used now primarily as a mitigating factor but not providing a
defense to the need to use reasonable alternative designs in avoiding
liability. But in any event, it seems that the scale of nanotechnology
is such that design defects would rarely be "open and obvious." Can
you think of any nanotech examples where that might be the case,
thereby affecting the total liability for producing products that have
such a design defect?

6. In some instances, the design of a product may be subject to some
safety regulation. The relevance of such regulations is asymmetric.
Violating the regulation is frequently sufficient to prove the design
is defective. But if the design complies with the regulation, that fact
merely acts as some evidence that the design is not defective—the
possibility is held out of proving the design to be defective with
additional evidence. How do you account for this asymmetry?

7. A consumer purchases a state-of-the art computer monitor from a
start-up company that uses nanoscale electrodes in the stimulation of
molecules to produce an extremely high-quality image. Because the
electrodes are defective, the consumer is electrocuted and has damages in the form of current and future medical care, lost earnings, and
so forth amounting to $35 million. The start-up company declares
bankruptcy. The specific manufacturer of the electrodes cannot be
identified, but there are only three producers of such electrodes in
the world. Can the consumer recover his damages? Do you see any
unfairness in your answer, either to the consumer or to the electrode
producers? Can you suggest a way to resolve the unfairness?

8. It sometimes seems that warnings have proliferated in recent years, so much so that their impact is lost. Do you think it is possible for a warning to be inadequate by having too much information or by being presented alongside too many other warnings? If so, can you articulate an effective way to determine when the competing needs of providing sufficient information without overdoing it are balanced to provide an adequate warning?

9. The last several years have seen a rise in "direct-to-consumer" advertising of products that are unavailable for direct purchase by consumers. For example, television commercials might tout the benefits that the use of nanotechnology provides in strengthening synthetic dental structures, which are available for purchase only by dentists. How does such advertising affect the learned-intermediary doctrine in evaluating warnings? Should a manufacturer be liable for injuries resulting from the product if its risks were adequately conveyed to the dentists who purchased the product but omitted from the direct-to-consumer advertising?

## 3. Warranty

Do manufacturers of nanotechnology products stand behind their goods?

The discussion of civil liability for nanotechnology products and treatments has so far focused on liability that arises from theories that are said to *"sound in tort."* What this means from a theoretical perspective is that the basis for liability arises from violation of a social duty that is generally applicable. The requirement that people behave reasonably under the circumstances is an example of such a general duty; it applies to everybody within a particular societal context and a violation of the duty gives rise to liability under a theory of negligence. Similarly, the requirement that commercial products be provided free of design defects is a duty that applies to everyone who provides commercial products, with a violation of that duty also giving rise to liability.

It is also possible for liability to arise from theories that *"sound in contract."* The fundamental difference between tort- and contract-based theories is that contracts are defined by agreements between individuals; the precise parameters of such agreements are not constrained by general duties but are instead circumscribed by the parties themselves. There are innumerable ways in which the existence of contracts involving nanotechnology can give rise to liability. Employment or consulting agreements involving work to be done by nanotech scientists and engineers, purchase-order agreements to buy a certain number of nanotech products, supply agreements to provide nanotech manufacturers with raw materials, intellectual-property licensing agreements of nanotech patents, and countless other contractual agreements can arise in nanotechnology industries. When a party to one of these agreements fails to perform its specified obligations—to make a payment, to supply a product by a certain date, to deliver a consulting report, and so on—it may be liable for damages that result from a contractual breach.

The law of contracts has many facets that address such issues as when a contract actually exists, what its terms are, when it needs to be in writing, what constitutes a breach, what the permissible scope of damages may be from a breach, and when a party may be excused from the contract. Of particular interest here is the law of warranty, which is a particular form of contract in which a seller of a product promises that the product will have certain characteristics in exchange for the money provided by the buyer who purchases the product. The existence of such warranties may provide an alternative basis for consumers of nanotechnology products to recover when the products do not perform as expected.

Although tort theories and contract theories have different conceptual underpinnings, there are some areas in which the resulting scope of liability very much overlaps. At the same time, there are also areas where contract theories provide a more natural basis for the discussion of liability. Consider, for example, a company that produces and sells blood-pressure sensors that incorporate nanoelectromechanical systems (NEMS) structures. A customer purchases one of these devices and it just doesn't work. The instructions for using it are followed correctly, but the display always reads "0/0." It would theoretically be possible to bring an action in tort alleging that there is a manufacturing defect with the product. But no one has been injured and no property has been damaged. The more natural recourse is based in contract.

Or consider a company that sells trousers advertised to include nanoparticles that make them much more stain resistant than trousers that use some kind of coating. It turns out that whenever customers spill things on the trousers, they stain in pretty much the same way as any other pair of trousers. The nanoparticle structure is completely ineffective in enhancing stain resistance. One mechanism for redress by a customer might be to assert that there is a design defect. But, again, no one has suffered any injury and no property has been damaged (except perhaps the trousers themselves). A contract-based approach would instead focus on the fact that the merchandise did not perform as advertised.

### i. Express Warranty

The most straightforward warranties are those that are made explicitly by the seller of a product—*"express warranties."*[25] These warranties frequently take the form of a written guarantee that sets forth what is being promised by the seller about the product. Such a written guarantee is often even captioned "warranty" and may be provided as a separate document sold with the product, be included in an instruction manual, printed on packaging, or conveyed in some other manner. But an express warranty need not be highlighted in this way. In general, any statement of fact made by the seller has the potential to be treated as an express warranty and to act as a promise by the seller that the product conforms to that statement. Warranties can often be made orally. And they can frequently be made without the seller even intending to create a warranty. Sources of express warranties therefore include not only the "Warranty" section of an owner's manual, but potentially any factual representation made elsewhere in the owner's manual, made in advertising for the product, or even made by the salesman on a showroom floor.

The principal issue that can be raised when discussing express warranties is whether the statement was a factual one. Consumers of products are expected to understand that sales pitches often include exaggerations of opinion, and such statements will not be held to act as promises about a particular product. For instance, consider a statement in a pamphlet sold with an antiwrinkle cream that the product "includes nanosomes that have been clinically proven to reduce the presence of lines around the eyes within six weeks when used as directed." This is a purely factual statement. It acts as an express warranty that clinical studies have been performed and that they provide evidence that crow's-feet can be reduced by using the product. If it turns out that no clinical studies were ever performed or that the results of studies that were performed clearly contradict the claim, there has been a breach of an express warranty.

Compare this with a statement made in a television advertisement for sunglasses that have a thin scratch-resistant polymer coating that includes nanosized particles: "These are simply the most stylish sunglasses on the market. Don't you want to look your very best?" All of this is exaggerated opinion. Perhaps others think the style of the glasses is ugly. Perhaps you don't actually look "your very best" when you wear them. The fact that your head shape and hair color do not suit the glasses does not result in a breach-of-warranty action because you do not look "your very best" when you wear the product.

These examples illustrate the difference between advertising "puffery" and factual statements fairly clearly. In other cases, it is not quite so obvious which side of the line a particular statement might fall on. Suppose you're looking for some new nonstick bakeware. A salesman starts pushing a particular brand, telling you that "this is the latest thing. It has a new 'nanotech' coating that is far superior to Teflon. It costs a little bit more, but trust me, it's definitely worth the extra money." You spend the extra money and a few weeks later read a report comparing various brands of nonstick bakeware. The costly nanotech bakeware performed about average and a little less well than most brands that use Teflon.

Is there a breach of an express warranty? Is the salesman's assertion that the nanotech coating is "far superior to Teflon" a representation of fact or mere puffing? It seems possible to argue either way, particularly if additional facts are introduced that suggest one interpretation over another. Legitimate disputes can and do arise over whether products have been warranted to have certain features, and such disputes will inevitably arise over nanotechnology products. This seems especially to be a concern with products whose features might be hyped in explaining how the nanotech aspect affects them.

Some statements might be construed as factual representations of the product that are not later borne out.

### ii.   Implied Warranty of Merchantability

It is also possible for warranties to arise by implication without any express statement being made by a seller. In its current form, this implied warranty is referred to as a warranty of *"merchantability,"* meaning fairly simply that the very fact that a seller accepts money for a product means that he is by implication guaranteeing that the product will be fit for its ordinary purposes.[26]

The pedigree of this doctrine is somewhat interesting. Almost everyone is familiar with the Latin phrase *caveat emptor,* "let the buyer beware," suggesting that the risk of a product defect rests with the buyer.[27] The fact that this phrase is universally quoted in Latin suggests that it enjoys a long history. And, indeed, this doctrine was the norm in American and English law for centuries. But in actual Roman times, the doctrine was rejected and a doctrine of *caveat venditor,* "let the seller beware," was adopted instead.[28] As applied in early Roman law, *caveat venditor* bears a strong resemblance to the current implied warranty of merchantability. If a purchaser of an item discovered a defect that made it unsuitable for its purpose or reduced its value, he could sue to annul the sale or for a price reduction.[29]

While it is possible for there to be some dispute over what is meant by "ordinary purposes," the application of the warranty is otherwise relatively straightforward. In fact, it has been noted that for a product not to be "fit for its ordinary purposes," it must contain some defect, either individually in the form of a manufacturing defect or as a more general characteristic of the product in the form of a design defect. As such, the theory of liability that underlies defects in strict liability is in some sense equivalent to the theory of liability that underlies breach of the warranty of merchantability. Although the theories "sound" in different legal doctrines, they converge on the expectations that they impose on the sellers of products.

### iii.   Implied Warranty of Fitness for a Particular Purpose

There is another implied warranty that is applicable much less commonly than the warranty of merchantability and may find some application to nanotechnology. The implied *"warranty of fitness for a particular purpose"*[30] extends the warranty of merchantability to instances when a buyer has a specific purpose in mind. The warranty is only created when the seller knows of this special purpose and the buyer relies on the seller's judgment in selecting the goods.

Suppose, for example, that a resident of Fargo, North Dakota, is looking to purchase a new set of tires and visits a tire shop in late October. She explains to the salesman that she would like new tires that will get her safely through the winter. The salesman recommends a new type of tire that includes a nanotech silica-based compound that has been shown in tests to improve tire grip to asphalt surfaces. It turns out that there is so much snow on the roads in Fargo that the new tires are no better than standard radial tires and that a better recommendation would have been to purchase conventional snow tires. In this example, there has been a breach of the implied warranty of fitness for a particular purpose because the buyer had a specific purpose—to use the tires in weather typical of winter in Fargo—that was communicated to the seller and because the buyer relied on the seller's judgment in making her selection.

The interaction between buyer and seller in these circumstances is such that there is frequently also an express warranty. For instance, the tire salesman might have said something to the effect of "These nanotech tires are the best for winter driving" as part of his sales pitch. It might seem that this is an entirely formal construct since it is the same performance of the product that will produce the breach of both the implied warranty of fitness for a particular purpose and of the express warranty.

### iv.  Disclaimer

But the importance of the coexistence of two warranties can be understood when considering the effect of disclaimers. Well aware of how warranties are created and particularly cognizant of the existence of implied warranties, most sellers attempt to disclaim them. Boilerplate language is included in almost every consumer contract indicating that (at least) no implied warranties are being made. To be sure, there are significant constraints on the form that the disclaimer can take, usually requiring that the disclaimer be made in a generally conspicuous manner. Most sellers comply with these requirements and the existence of a disclaimer still rarely dissuades most consumers from proceeding with a purchase.

But there is a marked difference between implied and express warranties that is manifested when discussing disclaimers. There seems to be a fundamental incompatibility between making an affirmative statement about a product and then retracting it in the form of a disclaimer when the buyer relies on it. For this reason, it is much more difficult to disclaim an express warranty. To the extent a disclaimer is possible, it is really intended only to better delineate the precise scope of the warranty. For instance, one area of nanotech consumer products gaining increasing attention is the use of hydrophilic nanometer scale elements in producing self-cleaning products. If a salesman indicates that this self-cleaning feature is fully warranted when pitching the sale, it is permissible to explain later that this express warranty is for some limited period of time, say three years. While arguably a form of disclaimer, it is considered acceptable because it is acting more to define the scope of the warranty than to circumvent it.

The question posed at the beginning of the section, "Do manufacturers of nanotechnology products stand behind their goods?" was intended to be rhetorical. Of course, virtually all manufacturers of nanotechnology products will claim that they stand behind their goods and intend for their customers to be satisfied with their products. But there are limits to what any manufacturer will tolerate in the form of consumer demands. The law of warranty provides a framework in which these limits can be defined. It is possible for some manufacturers not to stand behind their goods at all, to sell them "as is," with no express warranty and with all implied warranties disclaimed. Others will be generous in providing express warranties, believing that the warranties will enhance sales. The law of warranty permits wide latitude among different sellers to take these different approaches.

### Discussion

1. The text notes that a product characteristic that results in a breach of the implied warranty of merchantability almost certainly also qualifies as a manufacturing or design defect. Yet, such defects give rise to liability in a "strict" fashion, while the warranty of merchantability can be readily disclaimed. How do you account for this?

2. *Statutes of Limitations.* All civil actions are subject to a *"statute of limitations."* This is a law that sets a maximum time period during which a party is permitted to initiate a legal proceeding. If the party waits too long, he loses the right to bring an action. One of the main reasons for having statutes of limitations is to permit parties who might be subject to legal actions to reach a state where the possibility no longer weighs on their minds; it becomes sufficiently long ago that it can essentially be forgotten. The event that starts the time period can differ under different circumstances. In some cases, it begins when a certain act is committed, even if the party is unaware that the act was committed. In other instances, it begins when the party learns that the act was committed.

   In what ways can statutes of limitations interact with the ability to bring a lawsuit under different legal doctrines for the same underlying facts? Consider the following examples, in which today's date is January 2, 2012, and the statutes of limitations (SOL) run from January 1, 2009.

   a. Tort SOL 3 years; warranty SOL 4 years.
   b. Tort SOL 4 years; warranty SOL 3 years.
   c. Tort SOL 3 years; warranty SOL 4 years; implied warranties properly disclaimed.
   d. Tort SOL 4 years; warranty SOL 3 years; implied warranties properly disclaimed.

   What legal options exist for a plaintiff under each scenario? Do any of your responses seem to be unfair? If so, why do you think the law tolerates such unfairness?

3. *Uniform Codes.* Because of the federal structure of the United States, there is the potential for significant variation to exist in the laws of

different jurisdictions. One effort to harmonize the laws of different jurisdictions has been led by the American Law Institute. Under its auspices, authoritative panels of attorneys, judges, and legal scholars develop sets of laws that are in some sense "ideal." While everyone recognizes that this is an unachievable goal, it is hoped that the *"uniform codes"* that are developed take account of various issues sufficiently well that most jurisdictions will enact them and bring consistency to laws among the different jurisdictions.

Many of these uniform codes have been very successful. The Uniform Commercial Code, for example, which provides the basis for the warranty discussion in this section, has been adopted in its essential form by all U.S. states.[31] A newer uniform code that has not yet been nearly as widely adopted is the Uniform Computer Information Transaction Act. This code is intended to govern contracts that "involve information," including the development and licensing of computer software. One implied warranty created by this act is an implied warranty of *"informational content."* The warranty is modeled after the warranty of fitness for a particular purpose when applied to goods. It provides an implied warranty of accuracy in "informational content" transmitted to a licensee that relies on a merchant to receive it.

Can you devise any hypothetical scenarios involving nanotechnology where such a warranty is breached? Research other implied warranties set forth by the Uniform Computer Information Transaction Act. What circumstances involving nanotechnology can you imagine where they would be implicated?

4. Another warranty that the Uniform Commercial Code provides is a warranty "against infringement." Under this implied warranty, a merchant who supplies goods to a buyer warrants that they will not infringe the rights of third parties. In view of the discussion of "patent thickets" (see the first chapter), how significant do you think the potential liability is for breaching this warranty? Can you think of any reason why a merchant might not want to disclaim this warranty? Besides disclaimers, what steps can merchants take to minimize the liability?

## 4. Class Actions

Your temptation is whetted when you yet again see the slick ad promoting a new tennis racquet. The deep voice rumbles: "Nanotechnology has created the most advanced tennis racquet the twenty-first century has to offer. Increased power. Faster shape restoration. More accuracy. It all adds to up to providing you with the deadliest serve technology has to offer." The promotion comes with brightly colored visuals of players using the racquet to

deliver unreturnable serves and highlighting the improved scores the players achieve.

After fretting about the high cost, you succumb and spring for the $299
retail price that comes with the racquet. The first several matches you play
with it convince you that it actually lives up to the hype. Your serves seem
more powerful and more accurate. Your overall game seems to improve.
But then one afternoon, as you swing the racquet overhand to deliver your
new precision serve, the racquet shatters explosively. Shards embed themselves in your face and you require emergency medical treatment. As you
are transported to the hospital, the deep advertisement voice intoning about
"deadly serves" plays over and over again in your delirious state.

After you are treated and recover, you consult with an attorney to see
what compensation you might be entitled to from the racquet manufacturer.
The attorney has seen exactly this happen before. Three or four others have
approached him with similar stories. "The pattern always seems to be the
same and your injuries are pretty similar to the kinds of injuries others have
suffered. But when we've approached the manufacturer, they've consistently
refused any form of settlement. I'm afraid the only option would be to litigate the matter."

He reviews your damages with you and they are fairly modest: the cost
of the racquet, some medical bills, a couple of days wages lost from work,
perhaps some pain and suffering over the time it took your face to heal—in
all, somewhere around $8000. Then he reviews the cost of litigation: attorney
fees, discovery, hiring expert witnesses, and so on—in all, somewhere around
$150,000. Your case is just too small for him to take on a contingency basis.
And there is no sense funding a $150,000 litigation just to recover $8000.

You return to your life, feeling angry and frustrated at how the legal system fails to provide any reasonable way for you to recover. Your injuries were
real; your expenses were legitimate. But the financial realities simply preclude
an effective way to bring your case. A few months later, you receive a telephone call from the attorney. "It seems that there have been others who purchased exploding tennis racquets," he explains. "Although the manufacturer
has now withdrawn the product from the market, we've identified a total of
seventy-two consumers just like yourself who suffered similar injuries." The
attorney goes on to explain that it might be possible for all seventy-two of you
to try your case collectively in a *class action*. There are some administrative
complexities that would raise the total cost of the litigation, with him now
providing an estimate of about $180,000. Shared among seventy-two plain-
tiffs, though, the cost to each of you would only be about $2500—pretty close
to the one-third fee that you would have paid on a contingency basis. On this
basis, the economic considerations seem much more reasonable.

### i.   *Class-Action Criteria*

The ability to participate in a class action is an important legal mechanism
for allowing actions with small individual recoveries to be pursued. It is easy

to see how the absence of such a mechanism would permit parties who create many small harms to avoid liability for them. It would simply be economically unfeasible to proceed with individual actions because the cost of the actions would be all out of proportion to the individual potential recoveries.

The mechanics of proceeding with a class action vary according to jurisdiction, although there are some general features that they have in common. Probably the most evident feature of a class action is that the number of parties on at least one side of the litigation is large. While there is no predetermined number of parties that triggers the use of a class-action procedure, it must generally be impracticable to proceed in some other fashion.

While class actions are frequently thought of in terms of the more typical pattern of a large number of plaintiffs joining together as a class and suing a corporation or other party, there is no requirement that the class must act as a plaintiff. It is equally possible for defendants to be organized into a class, although much less common. In some especially rare circumstances, parties on both sides of the litigation are organized into classes in what is referred to as a *"bilateral class action"* (Figure III.6). For example, one could imagine a scenario in which nanotubes are found to cause respiratory illnesses on a wide scale and without being particularly limited to the types of products that use them. It is conceivable that a bilateral class action might be filed. On the plaintiff side would be a class of users of products that incorporate

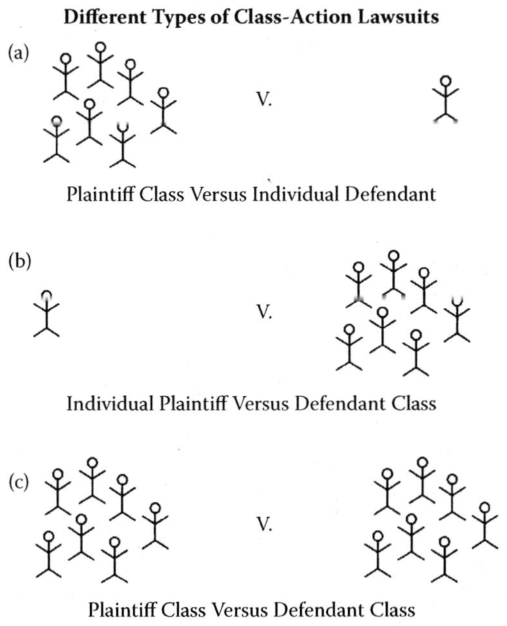

**Different Types of Class-Action Lawsuits**

(a) V.

Plaintiff Class Versus Individual Defendant

(b) V.

Individual Plaintiff Versus Defendant Class

(c) V.

Plaintiff Class Versus Defendant Class

**FIGURE III.6**
Class actions may be formed with classes on either or both of the plaintiff and defendant sides of an action.

nanotubes. And on the defendant side would be a class of product producers that manufactured those products.

Irrespective of which side of the litigation includes parties organized as a class, a small number of those parties—frequently just one—is designated as the representative(s) of the class. In order for a class action to proceed, the representative(s) must be situated so that the arguments available to them in litigating the dispute are typical of the arguments available to every member of the class. The objective, after all, is to simplify the consideration of issues for all class members when the legal and factual issues are generally similar. Thus, another requirement of the representative(s) is that they will participate in the action in a manner that reflects the interests of the class as a whole, and not parochially emphasize their own interests to the exclusion of those of other class members.

## ii. Certification

If all the threshold criteria for a class action are met—numerous members with common issues who will be adequately represented by a typical class member—then the class may be *certified* by a court. This means that the court has sanctioned the request to proceed with the matter on a class-action basis, recognizing a particular class that has the requisite characteristics. Depending on the jurisdiction, other considerations may also factor into whether to certify a class. For instance, one consideration might be whether there is a risk of actions involving different plaintiffs resulting in inconsistent duties being imposed on a common defendant.

Certification of a class can be highly contentious, and there can be significant opposition by the other party to the way the class is defined.[32] Such disputes arise because of the way modern class actions tend to be implemented. In the narrative at the beginning of this section, the parties to be included in the class were aware of the litigation strategy before it was adopted and it was clear how legal expenses would be paid. While class actions can still arise in this way in certain jurisdictions, the perspective of most consumers increasingly sees a different pattern. It frequently seems to many consumers that they first learn of a class action when they receive a coupon in the mail for some nominal amount and read in the newspaper that attorneys have been awarded a multimillion-dollar fee.

This type of scenario arises because class actions are increasingly structured as "opt-out" actions.[33] In such a structure, any person who satisfies the conditions set in a definition of a class is automatically considered to be part of the class. For instance, a class might be defined as all persons who purchased a NanoRacquet5000 within the state of California between April 2005 and January 2007. If someone who falls within the definition does not want to be bound by the class action—because she wants to preserve her rights to sue independently of the class—she must affirmatively opt out of the class by notifying the court. In some instances, an individual party might meet the class definition, but believe that she has greater damages or is otherwise able to recover a greater amount by acting independently.

Because the way a class is defined has the potential to impact the rights of individuals in a significant way, there is always a notification requirement associated with the certification process. In most cases, every person comprised by the class definition must be notified by mail of the action and informed of the procedure to opt out. Sometimes broad public notifications may also be provided in the form of newspaper or other widely publicized advertisements. Ultimately, the details of how notification is to be carried out are discretionary with the court. The use of these notification mechanisms means that the impression that some have of learning about a class action only after the fact are almost certainly exaggerated. The notifications are generally effective, with most people receiving some notice of their inclusion in a class, even if they disregard it. In almost all instances, parties notified of a class action in this way take no action, thereby proceeding as participants in the class.[34]

The debate over the definition of the class is, in reality, a debate over how many people to include in the class. But it is not always the case that the party opposing the class prefers a narrower definition. To be sure, by defining the class narrowly, the immediate financial exposure is reduced. But it also leaves a larger group of people outside the class having retained their rights to bring individual actions. Defining the class broadly may be advantageous to the opposing party because it provides greater certainty to the overall financial exposure by extinguishing most of the outstanding liability.

In principle, once the class has been certified, much of the manner in which litigation proceeds is the same as in any other litigation: parties engage in a process of evidentiary discovery to ascertain the strength of each other's positions, a trial is held where evidence is presented, and a judgment is rendered. One acknowledged difference is that there is even greater court supervision of details of the litigation than in non-class-action lawsuits. In practice, the number of class actions that proceed to trial is very small. One study found that a trial is begun in only 4 percent of class actions that are filed.[35]

### iii. Class-Action Settlements

While a societal perspective favors settlement in virtually all civil litigation, there are special issues that surround settlement of class actions. The norm in civil litigation is for courts to refrain from meddling in reaching a settlement—the parties are ordinarily free to reach any agreement they wish without needing to seek approval from the court. But because of the representative nature of class actions, court approval is required of any settlement.

The process by which settlement is approved bears some similarity to the process by which the class was initially certified. A proposed settlement devised by the attorneys representing the class and the opposing party is presented to the court for preliminary approval. If the court accepts the proposal, a settlement notification process is initiated that is similar to the certification notification. Every member of the class must be notified of the proposed settlement and is given a second opportunity to opt out of the class based on this settlement information. The opt-out rate at this stage is generally higher than the opt-out rate at the certification stage.[36] Each of the class members may also be given an opportunity to be heard by the court regarding the fairness of the settlement. The class members may present evidence—sometimes merely in the form of their own testimony— at a hearing specifically convened to assess the fairness of the settlement proposal.

In most cases, the case essentially concludes with approval of a settlement. In instances where there is no settlement, the case proceeds much like conventional cases with a trial at which the parties are able to present evidence to support their respective positions. One aspect that is characteristic of a class action is the focus throughout the action—including the trial—on the representative party, with the specific factual issues that apply to the representative party dominating over the individual factual characteristics of other members of the class. It is this focus during the actual consideration of the case that makes it important for the factual underpinning of the arguments available to the representative to be fairly typical of those available to all the members of the class.

The trial of a case presents a number of procedural options that are truncated if the case settles. Probably the most evident is the availability of appeal. Either party—the class or the party opposing the class—almost always has the option of appealing a final judgment by a judge or jury. This has the potential of significantly lengthening the total time to conclude the action as all options are exhausted in a way that is not an issue when there is a settlement. But if the plaintiff ultimately prevails, the end result is similar to having reached a settlement in that the defendant is then obliged to provide the resulting recovery to the plaintiffs.

### iv.  *Underlying Causes of Action*

Some of the class-action lawsuits that have the highest profile are those that involve product-liability issues. There is every reason to expect that many class actions involving nanotechnology will be of this type. Indeed, the narrative at the beginning this sections motivated the discussion by attempting to illustrate just such a class action. But it is important to recognize that class actions are not conceptually limited to any particular type of legal action. In principle, any type of complaint that finds legal support may be brought as a class action if the criteria discussed earlier are satisfied.

This accordingly includes not only actions for design defects, warning defects, and manufacturing defects, but also for actions that involve other torts like negligence, nuisance, or even intentional torts. And there is no restriction to tort theories. Class actions may be brought under breach of warranty or other contract-based theories if appropriate. In recent years, there has been an increase in the number of class actions brought under fraud theories, particularly involving securities fraud.

Indeed, one of the earliest class-action lawsuits ever raised against a nanotechnology company alleged a violation of the Securities Exchange Act. The facts as they are alleged in this case are interesting because they illustrate how intellectual property can indirectly result in potential liability in a class-action setting. They also illustrate what is becoming a common type of allegation that nanotechnology businesses need to be aware of.

The company involved in this case is NVE Corp., which describes itself as having helped to pioneer "*spintronics.*" In the same way that now-conventional electronics makes use of the charge states of electrons in transmitting information, spintronics makes use of the intrinsic angular-momentum ("spin") states of electrons. The idea that electrons have an intrinsic angular momentum was one of the first discoveries to result from the development of quantum mechanics in the early twentieth century. Referring to this property as *spin* allows a visualization of an electron as a solid sphere that rotates about its own axis in a manner similar to the Earth's daily rotation. It turns out from the application of quantum mechanics that an electron can have two intrinsic angular-momentum states, visualized as the electron spinning in either direction about its axis (Figure III.7).

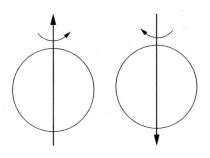

**FIGURE III.7**
An illustration of the concept of electron spin, a form of angular momentum intrinsic to the structure of an electron itself. Electron spin can be in one of two states: up or down.

The existence of two electron spin states provides a somewhat convenient mechanism for a binary representation of information. And information can be conveyed by the movement of spin in a manner analogous to carrying information electronically by the movement of charge. Spin is susceptible to change by the application of magnetic fields, which are already commonly used in magnetic storage devices.[37] It is perhaps not too surprising, then, that one of the technologies that finds particular application for spintronics is in memory structures. A particular area of focus by NVE has been in magnetic random-access memory (MRAM) structures.

The allegations that form the basis of the class action are that NVE issued a number of press releases that misrepresented the technical state of MRAM in order to inflate its stock price. It is true that NVE's stock price rose rapidly from a low of $5.86 on May 22, 2003, to a high of $66.80 on January 16, 2004. And its stock price dropped after one of its licensees, Cypress Semiconductor, sold its stake in the company and made comments on February 14, 2005, questioning the economic feasibility of the technology. The manner in which the class is defined in this action illustrates the way in which the definition attempts to be tied closely with the underlying allegations: the class is defined as those who purchased common stock on the open market between May 22, 2003—the date of the stock's low—and February 11, 2005—the business day immediately preceding the Cypress Semiconductor announcement.[38]

But the truth of the allegations and their merit in establishing that there was a violation of securities law is not of particular concern here. Instead, this example is raised because of what it can illustrate about the scope of liability that may be faced by nanotech companies in real-world circumstances. Civil liability does not arise merely from harms that the nanotech products themselves cause, but may arise from more indirect issues. In the NVE case, potential liability has resulted from the manner in which the technical field as a whole was characterized.

Other indirect ways in which liability might arise are from claims made about the effectiveness of nanotech products. Consider, for instance, a class-action lawsuit that is being closely watched by manufacturers of sunscreens that incorporate nanoscale $TiO_2$ particles. On March 30, 2006, lawsuits were filed in the Los Angeles Superior Court[39] alleging that various representations about the effectiveness of conventional sunscreens were known by the manufacturers to be inaccurate and created a false sense of security in consumers who used them. The assertions are that conventional sunscreens, even when used as directed by the manufacturers, do not offer the protection against skin damage by ultraviolet light that the manufacturers claim. Again, the truth of these allegations is not particularly relevant here. Instead, this case illustrates the care that manufacturers of nanotech products—not just of sunscreens—are advised to exercise when advertising their products.

Still other indirect ways in which liability might arise center on the pricing of nanotech products. Allegations may be made that the pricing models are anticompetitive, that they misuse monopolistic market power, or are in some other way improper. In some respects, these examples seem to be on the periphery of how liability might arise because they are not directed specifically at the way harms result from product usage. Even the sunscreen case, while obliquely relying on the fact that users of sunscreens are still at risk for developing serious cancers, is based not on a harm caused by the sunscreen itself but on the way its efficacy was characterized to consumers. In reality, these types of cases are not at all uncommon. And when they are filed, it is almost routine for them to be filed as class actions.

## v.  *The Regulatory Role of Class Actions*

In recent years, both the number and size of class-action lawsuits have increased significantly and rapidly. Without considering the underlying reasons for this trend, it is worth commenting on its effect since it defines a legal atmosphere in which nanotech companies will operate. Almost certainly, the most significant aspect of this atmosphere is that the exposure to liability is notably greater today than it was even a relatively short time ago. While it has always been the case that a company might be sued by an individual for any of the various kinds of actions mentioned earlier, the ease with which a class action may be initiated has the effect of magnifying the total liability in a very immediate way. Issues that by themselves are worth only small amounts of money to individuals suddenly expose companies to multimillion-dollar levels of liability.

Many lawsuits that would never have been pursued in an earlier litigation climate are now almost routinely filed. While this may be onerous to companies, it is not necessarily undesirable from a policy perspective. The fact is that regulatory agencies are not sufficiently well funded to police all the various regulations that they promulgate. And there is undoubtedly political resistance to governments using taxation as a mechanism to provide greater funding. The availability of class actions acts in many ways as a private form of regulatory enforcement. In cases where a nanotech company is not adhering to a promulgated regulation, it opens itself up not only to a relatively unlikely action by the regulatory agency but also to a much more likely class-action litigation.

A climate in which there are relatively frequent class actions also results in a widespread criticism that the real beneficiaries of such actions are not those who have been harmed, but instead the attorneys who try the actions. It is not uncommon to hear about class members receiving coupons that have relatively small value while the attorneys who brokered that settlement receive millions of dollars in fees. While there is some truth that this pattern is prevalent, the public perception is probably inaccurate; there is a wide spectrum of settlement values and many provide class members with amounts that are far from nominal.[40]

### vi. Attorney Fees

But in light of this widespread perception, it is perhaps worth noting how attorney fees are determined in class actions. There are essentially two techniques, both of which are variations on conventional arrangements for determining attorney fees. In conventional litigation actions, where a single party hires an attorney or law firm to represent him, payment is either time based or contingency based.[41] Time-based fees are calculated straightforwardly by multiplying an hourly rate of an attorney by the number of hours spent on a particular matter. While there is considerable variation in the hourly rates charged by attorneys, they are generally acknowledged to be high, with the time of the most expensive attorneys now being charged at rates in excess of $1000 per hour.

Such rates are beyond the means of many potential litigants. Contingency-based fees provide an alternative for plaintiffs. In such arrangements, an attorney agrees to represent a plaintiff in exchange for a portion of the total recovery. While the size of the portion varies, it is typically in the neighborhood of one-third of the recovery. This fraction has generally been found to provide reasonable compensation to the attorney, in light of the total amount of work required for the representation and in light of the risk the attorney takes that there will be no recovery.

In most plaintiff class actions, the attorneys representing the class are paid from a *common fund* that represents any financial benefit that accrues to the class.[42] As such, the payment of attorney fees in these types of class actions is a form of contingency-based fee since the attorneys are paid nothing if the class fails to recover. But the actual amount may be calculated in different ways in different cases. In some instances, it is calculated as a percentage of the common fund in a manner very analogous to conventional contingency-based fees. The percentage tends to be a little bit lower that the conventional one-third, usually being closer to one-quarter, but there is again considerable variation in the precise percentage used.

The other technique is referred to as a *"lodestar"* fee and results in a certain hybrid of the fee-based and contingency-based fee structures of conventional cases. A lodestar fee, $F$, is calculated as the product of an hourly-based fee with a lodestar factor, $f$:

$$F = f \sum_i t_i R_i$$

The hourly-based fee is the sum of the time, $t_i$, spent by each attorney multiplied by that attorney's hourly rate, $R_i$. Multiplication by the lodestar factor is intended to reflect the existence of risk undertaken by the attorneys. Because the fee is drawn from the common fund, it is inherently limited by the total size of that fund—if there is no recovery for the plaintiff class, there are no funds to pay any attorney fee. Lodestar factors can also vary significantly and account for an assessment by a judge of such factors as the complexity of the litigation, the quality of the legal services, and the level of risk of no recovery. While it is unusual for lodestar factors to be less than 1.0, they can sometimes be; usually, lodestar factors are in the range of 1.0–4.0.[43] The intent of both methods is to provide attorney fees that are reasonable. And most studies confirm that the two methods generally result in similar kinds of fees being awarded to attorneys involved in class actions.

All of these various issues surrounding class actions add up to an environment that can present significant exposure to liability for nanotechnology companies. The class-action mechanism has the ability to act like a lens in magnifying issues that are small individually into liabilities that are large in the aggregate. The manner in which attorneys are compensated for bringing class actions acts as an incentive for them to identify issues that are potentially most affected by this aggregation when viewed through the class-action lens. Companies formed around nascent technologies like nanotechnology are among the most susceptible to creating these kinds of issues. Even with rigorous testing, there are many more ways in which the interaction of their products with the broad consumer marketplace may develop unexpected concerns when compared with more mature technologies.

Benjamin Franklin once observed that "a small leak can sink a great ship." This is an apt metaphor for the effect that class-action litigation can have on a nanotechnology corporation. To the extent it is possible, the key to minimizing liability seems to be to pay attention to even the small leaks that will inevitably arise in the form of an endless variety of product defects.

## *Discussion*

1. The second discussion topic of the "Negligence" section considered the development of equitable principles and the courts of chancery. Class actions as they are currently implemented are an outgrowth of one such equitable principle. More traditionally, parties to a lawsuit were expected to participate directly in the lawsuit. In cases where multiple parties were affected by the same issues, their separate lawsuits could be combined in processes that include *"joinder"* and *"consolidation."* These procedures differ in that joinder is used when the issues in the different cases arise from the same transaction and consolidation is used in other circumstances where there are still common questions of law or fact. These processes are still used when the number of parties is manageably small and they provide a mechanism to evaluate evidence and consider the issues involved more efficiently.

    It was the flexibility embraced by the system of equity that first permitted rights to be asserted in a more representative way through a *"bill of peace."* Such bills were first recognized by the English courts of chancery in the seventeenth century and could be "filed when a person [had] a right which may be controverted by various persons, at different times, and by different actions."[44] One widely cited example of the use of a bill of peace is the 1681 case of *How v. Tenants of Bromsgrove*, 1 Vern. 22, 23 Eng. Rep. 277 (1681). This case arose in response to a nobleman appropriating village land in Bromsgrove to be used as a hunting ground. Tenants challenged the appropriation and were permitted to proceed in a representative capacity to avoid the need to relitigate the same issue repeatedly with respect to each tenant and to avoid the possibility of inconsistent results.

    One difference between the bill of peace and the more modern form of class actions is the way in which parties become members of the class. While class actions currently tend to use an "opt-out" mechanism like that described in the text where parties automatically become class members unless they take some affirmative action to avoid doing so, the bill of peace was structured in a manner that resembles an "opt-in" mechanism. Such opt-in mechanisms are still available in some class actions, but are increasingly uncommon. They operate by requiring a party to take some affirmative action in order to become a member of the class, with the default otherwise being exclusion from the class.

What advantages do you see to opt-in mechanisms? Why do you think the trend is away from opt-in mechanisms to opt-out mechanisms?

2. Even when a class meets the requirements of commonality by presenting similar issues among the different members, it is possible for there still to be some internal complexity among the issues within the class. Although uncommon, it is then possible to define subclasses, with each subclass independently meeting the requirements for certification as a class. What advantages can you see with implementing such a hierarchical structure? What disadvantages do you see? Can you construct an example involving a nanotechnology product where you think the use of subclasses provides an advantage?

3. The text presents an argument that the principal value of class actions is in providing a pseudoregulatory-enforcement function. Under this view, the recovery by the injured parties is largely incidental since the focus is more on punishment of a defendant. Do you agree with this view? Or do you think that a structure that produces such huge potential rewards for attorneys instead acts to motivate frivolous actions based on harms of so minor a nature that even the "injured" parties have deemed them inconsequential? Can you propose any structure that would keep the benefits of the pseudoregulatory nature of class actions without encouraging frivolous litigation?

4. One significant difference in litigation that occurs in the United States from many other countries involves which party pays attorney fees. In many countries, it is the party that loses the litigation who pays the attorney fees for both parties; in the United States, each party pays its own attorney fees except in especially egregious circumstances. Which approach reflects the sounder public policy? Do you think having the losing party pay the fees of the winning party is effective in discouraging frivolous litigation? Does such a rule act too strongly to discourage even legitimate litigation? Which rule is more sensible when the parties have very different financial resources, such as when a small group of individuals sues a large corporation for a harm they suffered? How do you think the dynamics of class-action litigation are affected by the different rules?

5. In February 2005, the United States enacted the "Class Action Fairness Act." Several of the provisions of this act were intended to bring more uniformity to class actions by expanding federal-level jurisdiction over them at the expense of state-level jurisdiction. Other provisions were directed at the perceived unfairness of coupon settlements. Specifically, the act requires that contingent attorney fees in coupon-settlement cases be based on the value of redeemed

coupons and not on issued coupons. In addition, mechanisms are provided to subject coupon settlements to greater scrutiny. What effect do you think this legislation has on the nature and frequency of settlements? Based on your answer, what effect do you think there is on the options available to defendants to limit their liability? Does this legislation increase or decrease the total potential liability to nanotech companies marketing products?

## 5. Nanotechnology Business Organizations

The conventional wisdom that 80–90 percent of all new companies fail is certainly overly pessimistic. And yet, the fact that this statistic is so widely accepted is a testament to the high level of risk that new companies nonetheless face. Early-stage companies are particularly subject to a variety of financial pressures, many of which are driven by the fact that they have yet to establish well-defined markets for their products. In fact, it is very frequently the case that their products are in more of a state of active development than are the products of more established companies. For at least the near future, this will continue to be the status of the majority of nanotechnology companies. While there are numerous sources for the financial pressures they face, a significant component of those pressures is the potential for liability that has been considered in the preceding sections.

The companion book in the Perspectives in Nanotechnology series on business issues by Michael Burke provides a much more detailed discussion of the various issues involved with starting and running a nanotechnology business. Of concern here is why nanotechnology businesses are primarily structured as corporations and what legal effects those kinds of organizations have. When it comes to organizing a business, there are, in fact, a number of different structures that could potentially be adopted. One of the considerations in determining which structure to use for an organization is how liability issues are addressed.

A perspective on the liability issues can be derived by considering a small group of people who want to develop and market some nanotechnology product—say a professor who has made a discovery that he thinks has commercial application and a few friends with experience in raising funds from investors and in business issues like those described in Michael Burke's book. If this group simply jumps in and begins to manufacture and sell products, they are personally liable for all of the risks associated with their activities. If damages are imposed as a result of a harm caused by one of the products, they may have to liquidate their own savings, sell their houses, cash in their retirement funds, and so forth in order to satisfy a judgment.

### i.  Corporations

What a corporation does is to provide a strong measure of protection against extending liability to the personal assets of the principal parties. This is a

protection that does not exist with the other major forms of business organization—sole proprietorships and partnerships. A corporation is considered
to be a separate legal entity—a "person"—that has rights and obligations
similar to those of *"natural persons,"* that is, of human beings. These rights
particularly include the ability to own property and the ability to enter into
contracts. Because the corporation is legally able to act like a separate person,
it can take actions that cause it to incur liability itself. In order to satisfy the
liability, the corporation draws on its own financial assets to make payment.
The general rule is that natural persons—even those that may be closely
related to the corporation—are not required to draw on their own assets to
satisfy the liability of the corporation. This is much like a situation where a
human being incurs liability for his actions: he is responsible for drawing
on his own assets in making payment and none of his friends is obligated
to assist.

But even with its status as a distinct person, the interests of a corporation
remain intimately related with the interests of certain human beings. A corporation is created by human beings when they deliver *"articles of incorporation"* that include certain particulars about the structure of the corporation
to a designated state official. The articles of incorporation typically identify
the name and address of the corporation, and set forth details of shares it
is authorized to issue. Shares may be issued to human beings, who then
have a financial stake as owners of the corporation. And decisions taken by
the corporation are manifested by decisions made by human beings: there
are directors who manage the corporation's business by setting policy and
appointing officers and the officers themselves act administratively to manage the activities of the corporation on a day-to-day basis. Because of their
positions, directors and officers are subject to certain duties that require them
to manage the affairs of the corporation in a reasonable way. As discussed in
the following, their failure to do so may give rise to a cause of action against
them personally.

### ii.   Piercing the Corporate Veil

There are circumstances under which the interests of human individuals
and of the corporation itself are so closely allied that an exception is made to
the liability protection. Reflecting its generally transparent nature, the barrier to liability that is provided by incorporation is universally referred to
as a *"corporate veil,"* and removal of that protection is referred to as *"lifting"*
or *"piercing"* the corporate veil. When the veil is pierced, the assets of the
shareholders may be reached in satisfying the liability. Commentators have
identified the issue of whether the corporate veil can be pierced as "the most
litigated issue in corporate law."[45]

While there are a number of factors that are considered in deciding whether
to pierce a corporate veil, all of these considerations are attempting to discern the answer to a basic underlying question that relates back to the very
definition of a corporation: Was the corporation acting as a separate person

or was it nothing more than an "alter ego" of the shareholders? If it is acting as an alter ego, then its actions will reflect the personal interests of the shareholders rather than the independent interests of the corporation. There is some overlap in these interests since the principal objective of the corporation is to make money for the shareholders, but there are indicia that suggest the corporation is not acting as an independent person.

The most telling indicia are financial. One clear sign that the corporation appears to be an alter ego of the shareholders is that there is intermingling of assets between the two, that is, where shareholders treat the corporation's assets as their own rather than simply collecting periodic payments of dividends. When corporate records are not maintained, when there are no regular meetings of the directors, and when there is a general failure to follow normal procedures in running the corporation, there is also a likelihood that a court will find the corporation not to be acting as an independent person. Another strong factor that may be considered is the capitalization of the corporation. Does it have sufficient funds to operate as a business? Or is it so undercapitalized that its only possible role is to try to shield the personal assets of the shareholders?

All of these are considerations that should be borne in mind when a nanotechnology company is operating. They are perhaps more important for relatively small corporations. One example might include an incorporated medical practice that has only a single physician who is administering nanotechnology-based treatments. It is certainly a sensible structure to use because of its potential in insulating the physician's personal assets from liability. But such structures are particularly susceptible to allegations that the corporation is merely the alter ego of the physician, particularly if structured with the physician as the only shareholder, the only officer, and with only intimates as directors.

It is perhaps worth noting that the effect of the corporate veil is, in any event, limited to actions that are ostensibly taken by the corporation itself. Actions taken by an individual thus need to be examined to determine whether they have been performed as an agent of the corporation or in that individual's personal capacity. Sometimes, the question can be a close one. Suppose, for instance, that a corporation is formed among a group of friends to develop a line of car-care products—cleaners, surface waxes, and so forth—that incorporate nanoscale particles. Duties are assigned according to the articles of incorporation to different parties, with David acting as the chief financial officer and all development and testing activities assigned to Margaret, the chief technology officer.

What is the result if David supplies a local car club with test products that turn out to cause respiratory ailments in dozens of car enthusiasts? Is the corporation liable, even though David was acting well outside his assigned scope of duties? There are arguments that can be made on both sides of this issue. And a follow-up question might be whether the corporate veil can be pierced. It is interesting to note the different possible results. If David is liable personally, only his assets can be sought. If the corporation is liable, but the corporate veil cannot be pierced, only the assets of the corporation can be sought. But if the corporation is liable and the veil can be pierced, the personal assets of the entire group of friends—David, Margaret, and the others—are at risk. The other officers of the corporation are likely to be forthcoming in arguing that David was acting on his own, outside his authority, and that it was he—not the corporation—that supplied the test products to the car club.

### iii.  Derivative Actions

There are other important situations in which directors and officers may be subject to liability, notably when they fail to fulfill their duties to the corporation. As a distinct person, it is the corporation that is being wronged by such failures and it is the corporation that should take action against the officers or directors. But a difficulty is created by the fact that these are the very people who control the activities of the corporation; they are not going to have the corporation sue them for their own wrongdoing. This dilemma is addressed legally by providing for *"shareholders' derivative actions."* The basic idea behind the shareholders' derivative action is that the owners of the corporation can compel the corporation to exercise its rights by taking action against the officers or directors.

The two duties that officers and directors owe to a corporation are a duty of loyalty and a duty of care. The duty of loyalty requires that the officers and directors not allow their personal interests to prevail over those of the corporation. The duty of care requires that they be diligent and prudent in managing the affairs of the corporation. When a shareholders' derivative action is filed, it will assert that at least one of these duties has been breached by the actions of the officers or directors. In many instances, the allegations are that both duties have been breached.

As might be expected, the specific nature of a shareholders' derivative action may vary significantly, depending on the number of shareholders involved. When the number of shareholders becomes fairly large, the manner in which the action is managed bears much resemblance to class actions, as discussed in some detail earlier. That is, each of the shareholders must be identified and kept informed of key developments in the action. When an action is initiated, they must have the opportunity to object to the action; if enough shareholders object, the action cannot proceed. When a settlement is proposed, they must similarly have the opportunity to object, and the settlement cannot be approved if the settlement is unacceptable to too many of the shareholders.

What is different about a shareholders' derivative action is that the shareholders are never actually parties to the action. It is the corporation that is suing the officers or directors at the instigation of the shareholders. With this structure, it is sometimes possible for shareholders to have the option of filing a derivative action against the officers or directors, or of filing a class action against the corporation itself. One factor in deciding which type of action to file might be related to the availability of evidence. For instance, suppose that shareholders of a nanotechnology company believe they were induced to purchase shares in the company because of misleading representations that violate certain securities laws. It may be easier to find proof that the corporation acted improperly in making the representations than to establish that some duty owed to the corporation was breached by the officers or directors.

This again illustrates a general flexibility that exists in finding parties liable. Just as one set of facts can be approached with different legal theories— negligence, strict liability, breach of warranty, and so forth—so too can a single set of facts permit actions against different parties. This was already evident in the discussion of different bases for liability, in which an injured party could elect to sue any of a number of different parties: the physician who uses a nanotech device, the manufacturer who made it, the engineers who designed it, the retailer who sold it, and so on.

The ability to prove the elements necessary to establish the cause of action against the different parties is certainly a factor in selecting which parties to sue. But another factor is the ability of the parties to pay. A common thread to all of the discussion in the last several sections on civil liability is reminiscent of the advice Deep Throat gave to Bob Woodward in his Watergate investigation: "Follow the money." As nanotechnology businesses become more successful and have greater financial resources, they will increasingly become targets for litigation. Legal theories that were once dismissed as having only a 20 percent chance of success will become viable as the potential financial payoff increases. The ability to organize businesses as corporations offers some assurance in how liability will be assigned, but at the same time raises still other possible legal theories that may be advanced.

### *Discussion*

1. The ability of a corporation to purchase property and to enter into contracts permits it to own other corporations. Sometimes complex structures may be created in which multiple corporations are wholly or partially owned subsidiaries of other corporations, and there may be multiple layers of such subsidiary structure, resulting in an intricate ownership tree (Figure III.8).

   If a harm results from an act attributable to a nanotechnology company that is one company in such a tree, how is liability apportioned? Is the total liability limited to the assets of the nanotechnology company that committed the act? Based on your understanding of veil piercing, explain the circumstances under which you believe the assets of a parent company are at risk. What about a sibling company? Research how veil-piercing doctrines actually apply under such circumstances. Do they differ from your analysis? Do you find any differences surprising?

2. Individuals who become officers of a corporation may rightly fear that the actions they take in performing their duties will subject them individually to liability. One way to reassure these individuals

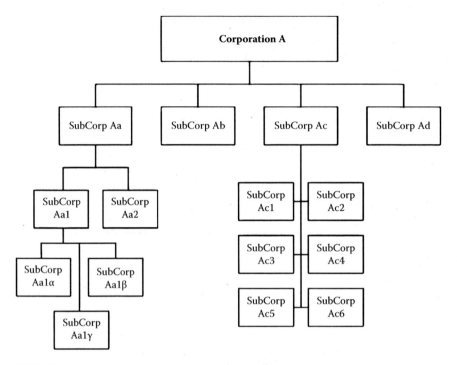

**FIGURE III.8**
An example of a complex structure that may complicate assigning liability for a harmful act.

is to have the corporation execute indemnification agreements so that the corporation will pay for any personal judgments rendered against officers for acts taken on behalf of the corporation. Assess the reasonableness of such agreements.

Now consider the following facts: A nanotechnology company is formed and indemnification agreements are executed with the officers. Shareholders believe that the officers have breached their fiduciary duties to the corporation and bring a shareholders' derivative action. After a year of litigation in which $1 million in attorney fees have been accumulated, the officers are found to have breached their duties and are liable to the company for a combined amount of $5 million. What payments are made—by whom and to whom—with this result? Do you have any further comments on the reasonableness of indemnity agreements between the corporation and officers personally?

## C.  Criminal Liability

Robert Stevens worked in Boca Raton, Florida, as a photo editor, a job that kept him indoors most days reviewing photographs that had been submitted to the tabloid newspaper that employed him. But he was also a keen out-doorsman whose passions were fishing and gardening. One day in late September, he took a three-day trip to North Carolina, where he indulged his outdoor passions. But he also developed a fever, nausea, and muscle aches. After returning home from the trip, he awoke in the night with a worsened fever and acting disoriented. His wife, Maureen, took him to the emergency department of a local hospital, where he was admitted. Mere hours later, he suffered a grand mal seizure, lost consciousness, and had a tube inserted down his trachea so that a ventilator could assist with his breathing. On his third day in the hospital, he died, having never regained consciousness. In addition to his wife, he was survived by three children.[46]

Robert Stevens died on October 5, 2001. An autopsy confirmed the diagnosis: he was the first person in almost a quarter century to have contracted inhalational anthrax in the United States. In the coming days, the pattern of his disease would be repeated multiple times, resulting in the deaths of an additional four people and in the sickening of seventeen others. It is likely that some time before his North Carolina trip, Robert Stevens opened an envelope that contained a fine powder laden with anthrax spores. He inhaled some of the powder and the disease took its natural course. It is unlikely that he ever made this connection since there were no reports of this biological attack at the time he would have opened the envelope. When other victims who had heard reports of prior attacks discovered powder in the envelopes addressed to them, they were in a better position to understand the nature of what had just happened to them. The person(s) responsible for these attacks has not been identified and the motivation for the attacks remains obscure.

The size of an anthrax spore is about 1000 nm, somewhat on the large end of a usual range cited for the size of typical nanoparticles. The ability to deliver deadly anthrax in so easy a fashion—simply by mailing an envelope stuffed with powder—serves as a compelling reminder of the ease with which synthetic nanoparticles might similarly be delivered. At this stage, there is no reason to suppose that nanoparticles have yet been fabricated that could impact the human body anywhere as destructively as anthrax spores, which are, after all, biological materials. But the ability to manipulate matter on the nanoscale unfortunately raises the specter of being able to fabricate particles that could interfere in a destructive way with critical biological functions and that could be delivered to human beings.

## 1. Crimes Using Nanotechnology

As an illustration of a nanotech crime, this analogy to the anthrax attacks of 2001 is, of course, an extreme one. At a minimum, those attacks were a form of mass murder, and at one point in time they raised concerns that they were part of a broader tactic of international terrorism. Thankfully, most crimes that involve nanotechnology structures will be much more mundane. The most common crimes involving nanotechnology will almost certainly be corporate crimes that are similar to crimes that are unfortunately already too familiar: in the name of increasing profit margins, some companies may deliberately truncate testing protocols before releasing products containing nanoparticles to the public; some companies might discard waste containing nanoparticles in ways prohibited by statutes or regulations; some companies might ignore laws governing the transportation of nanotechnology products in ways that release potentially damaging nanoparticles into the broader environment; and some companies may deliberately fabricate products that include nanostructures they know to present certain health risks because the incidental benefits of such structures to their products are so great.

This is, of course, nowhere near a complete list of the kinds of crimes that may be committed using nanotechnology. Indeed, it would be somewhat reckless to attempt to list all the different kinds of crimes that might involve nanotechnology. Crime is just one of many areas in which the capacity of humans to demonstrate ingenuity is evident. It is simply a consequence of the breadth of human nature that if there is a way to gain some advantage using nanotechnology, whether it be lawful or not, someone is eventually likely to try.

The gamut of crimes that can potentially be committed using some form of nanotechnology is thus as varied as the spectrum of criminal behavior itself. For example, while violation of crimes that have a regulatory character are designed broadly to protect a populace and the environment in which it lives, other nanotechnology crimes may well be directed more specifically at individuals. Consider the class of crimes that are referred to as being *against the person*. Such crimes generally include the unjustified infliction of some form of bodily harm against someone, whether that harm be a minor harm such as may result from an assault or a grievous harm that results in death.

Murders, batteries, robberies, and other such crimes have the potential to be committed using any of a variety of tools fabricated using nanotechnology techniques. Perhaps the most apparent current research that may provide nanotechnology-based weapons are studies of nanoenergetic materials. This research investigates the ability to use certain nanoscale formulations in producing highly energetic reactions that may be embodied in more sophisticated bombs and other types of weaponry. Already, there is promising research that demonstrates the ability to use metastable intermolecular composites, sol-gels, and other structures in generating such reactions. In other examples, the very techniques that permit physicians to cure patients have the potential to be misused in causing harm to others—using nanotech methods developed for drug delivery that are twisted into methods for delivering very precisely targeted poisons, for example.

Another class of crimes is those that are *"against property."* These crimes frequently take the form of thefts, with larcenies, burglaries, and other such classifications distinguishing among the circumstances of the theft and the value of the property taken. Nanotechnology has the potential to provide new techniques for committing thefts, particularly in providing techniques for gaining access to secure areas. Materials and devices may be developed that can attack security systems, and the possibility may even exist to produce biometric spoofs so precise at nanoscales that biometric security detectors will be circumvented. Indeed, any of the many military applications that have been identified for nanotechnology, from lightweight protective armor to more effective hand-held nanoenergetic weaponry, may be adopted by criminals in committing crimes against persons or against property.

Nanotechnology even has the capacity to increase the sophistication of a variety of forms of fraud beyond the use of nanotech-based coatings in producing more realistic biometric spoofs. Fraud techniques may be applied just as well in spintronics-based environments as they are currently applied in electronics-based environments. And if proposals to produce effective biological-electronic interfaces are realized with nanotechnology developments, a whole new sphere of techniques for committing fraud will result.

The reasons that nanotechnologies may be attractive to certain people who commit crimes include the fact that they expand the repertoire of available techniques, particularly into areas where effective countermeasures are technologically lagging. But an even greater attraction may be the greater difficulty posed for detection. Consider, for instance, a person who uses nanotech drug-delivery techniques to apply a very targeted poison in committing a murder. The very benefit of these techniques that allows cancer chemotherapy treatments to target tumors without widespread poisoning of a patient allows toxic drugs to be delivered very precisely to critical organs. A forensic analysis of the body in such cases does not reveal telltale traces of the poison throughout the body and may be very much more difficult to detect. There is a much greater possibility that the correct cause of death will not be possible to ascertain, presenting obstacles not only to investigations of the murder but in developing evidence that could be presented at a trial. Similar motivations of avoiding detection may contribute to the use of nanotech-based surveillance techniques in ways that are outside legal sanction.

## 2. Definition of Crimes

What is useful to recognize about these examples of nanotechnology-based crimes is that there is very little need to change the definitions of particular crimes to accommodate how nanotechnology is used in their commission. In the abstract, definitions of crimes are more about the motivations and consequences of the people who commit them than they are about the instruments that are used. One thing that the various examples of crimes mentioned earlier have in common, and that distinguishes them from the forms of civil liability discussed previously, is that they involve a deliberateness on the part of someone to cause some type of harm.[47]

This deliberateness is a manifestation of the principle of crimes. More generally, the specific elements that define individual crimes may be grouped into five kinds of principles: *mens rea*, *actus reus*, concurrence, causation, and harm. The last two of these are applied to criminal liability in a manner similar to the application of causation and damages in a civil-liability context. It is the interplay of the other three principles of criminal law that are of most interest in defining the scope of criminal liability.

"*Mens rea*" is a Latin phrase that means "guilty mind." It refers generally to the fact that to be found liable for (most) crimes, a party must have had an intent to commit the crime. The law recognizes that there are different levels of intent that a person might have, and this is often reflected in the existence of a variety of "degrees" of a crime. Perhaps the most illustrative example of how *mens rea* may affect liability is provided with homicide, although there are many crimes that have different levels of liability depending on the mental state of the perpetrator.

While all forms of homicide involve the killing of another person, the different mental states that the killer might have bear on how seriously society views the crime. The most serious forms of homicide are "murders," which may include different degrees. A "first-degree" murder is usually one committed with both "malice"—in law meaning without just cause or excuse—and premeditation. If there is malice but no premeditation, the murder is of the second degree. At a less serious level are manslaughters, with voluntary manslaughter being performed without malice, although the killing is still carried out intentionally; this usually arises when a killing is performed in the "heat of passion." Involuntary manslaughter is a killing performed without malice nor intentionally, but instead results from recklessness or negligence on the part of the killer.

This is not intended to be a treatise on homicide, but rather to illustrate the more general variations that can result from a person's mental state in evaluating potentially criminal behavior. For there to be a crime, the law also requires that there be an "*actus reus*," which is a Latin phrase that means "guilty act." Characteristics of the action taken by a person in committing a crime can also affect which crime (or degree of a crime) has been committed. For example, there are a variety of theft crimes that may have a similar or identical *mens rea*—to deprive a person of his property—but which differ according to the actions taken by the perpetrator in doing so. "Guilty acts" that involve weapons may result in more serious crimes that do not, particularly if the weapon is a firearm.

The principle of "concurrence" further requires that there be a relationship between the *mens rea* and the *actus reus*, generally that they occur at the same time. To paraphrase a typical illustration of this requirement, suppose a researcher is studying metastable intermolecular composites. These formulations are prepared as mixtures of nanoscale powders that exhibit very efficient exothermic reactions. Unlike conventional chemical energetics that output energy through intermolecular interactions, such powders undergo intramolecular reactions that can produce more highly energetic output.[48] As intimated earlier, these formulations may be used in the production of nanotechnology-based weaponry.

This particular researcher is hopelessly insecure about a particular rival and has often fantasized about taking a knife and brutally stabbing him to death. He is also habitually careless in failing to secure the site where he performs tests of nanoenergetic materials. During one particular test, a person wanders into the site and is instantly killed by the force of a nanoenergetic explosion. Horrified, the researcher rushes to investigate and discovers that the person killed is his rival. His mood quickly changes and he quietly celebrates the death of his competitor.

What crimes has the researcher committed? Perhaps involuntary manslaughter since his reckless testing protocols resulted in a person's death. But certainly not any form of murder. This is true even though he has apparently formulated an intent to kill that particular person and even though that person died as a result of an action by the researcher. There is no murder because there is no concurrence between the *mens rea* and the *actus reus*.

The requirement that there be a concurrence between the *mens rea* and *actus reus* also manifests itself in other ways. For example, a person cannot be found guilty of a crime merely because of his thoughts, no matter how twisted or heinous. The researcher discussed in the example is guilty of no crime by fantasizing about the killing of his rival. It does not matter how detailed his plotting was—he may have decided which knife he would use for the killing, the time and place it would occur, how he would dispose of the body and weapon, and how to construct his alibi. But in the absence of an *actus reus*—some guilty act—there is no crime.

This type of planning does not even make the researcher guilty of an "*attempt*" crime. The existence of such crimes provides at least a twofold purpose. First, they embrace the common-sense recognition that a person should not escape conviction for a crime merely by being unsuccessful in completing it. The societal motivations in preventing certain types of behavior extend as much to unsuccessful attempts to engage in that behavior as to successful attempts. Second, the existence of attempt crimes provides a basis for law-enforcement personnel to interfere with the commission of a crime and to disrupt it before its completion without jeopardizing their ability to prosecute the perpetrator.

The punishment for attempt crimes can be as severe as it is for the underlying crime itself: people have been executed for attempted murders and some statutes in the United States currently permit the same sentences to be delivered for murders and attempted murders alike. To be sure, it is often the case that those convicted of attempt crimes incur lesser punishments than if they were to have completed the crime. At least sometimes this reflects the fact that the failure to complete the crime was a consequence of some legitimate mitigating circumstance that caused the perpetrator to hesitate so that the reduced punishment is a result of the application of discretionary sentencing factors.

But what is of particular relevance is that there is no abatement of the requirement of an *actus reus* for attempt crimes. "Mere preparations" to commit a crime are not considered to satisfy the *actus reus* requirement. Instead some "substantial step" must have been taken toward completion of the underlying crime. The dividing line between preparations and a step toward completion of a crime is an exceedingly fine one. Deciding on which side of the line a particular defendant stands is a highly factual inquiry that carefully considers all of the circumstances surrounding the act—and how they are similar or different to circumstances considered in precedential rulings—to make a decision.

## 3. Corporate Criminal Liability

With relatively few exceptions, the principles described in the previous section apply to establishing that a crime has been committed and for determining liability for the crime. Of particular relevance to nanotechnology, though, is the fact that many crimes—probably most—involving such technology are unlikely to be committed in a completely individual capacity. Instead, the most likely scenarios are those in which employees of corporations will commit crimes that are related to the activities of the corporations, raising the possibility that the corporations themselves will have liability for the criminal action.[49]

Furthermore, nanotechnology is being developed in an environment in which criminal liability for corporations is being addressed much more aggressively than in the past. Up until 1984, the fines that could be incurred by a corporation for its criminal liability were identical to those that could be imposed upon individuals. The result was that 60 percent of the fines incurred by corporations for violating federal criminal statutes were less than $10,000.[50] A very different scenario currently exists and is a result of a number of developments that began with passage of the Sentencing Reform Act in 1984, which was part of a larger package of legislation aimed at reforming federal criminal law, the "Comprehensive Crime Control Act of 1984."

The Sentencing Reform Act sought to implement a philosophy of determinism in issuing criminal sentences generally. This was achieved through the institution of sentencing guidelines and the creation of a Sentencing Commission charged with periodically refining the sentencing guidelines after studying their effects. Specific sentencing guidelines for corporations went into effect in 1991 and have since been subject to certain amendments. Of particular relevance to corporations, the act also provided new mechanisms for dictating aspects of corporate behavior through a doctrine of *"corporate probation"* that could be imposed in addition to criminal fines.[51] This doctrine is now frequently used to order corporations to publicize their wrongdoings, adopt remedial compliance programs, and to take other similar actions— forms of punishment that were unheard of twenty-five years ago.

The effects of these reforms have been significant, resulting in dramatic increases in criminal fines and other sanctions imposed on corporations that have been found to have engaged in criminal behavior. And this trend of increasing not only the magnitude of punishments on corporations, but also the variety of the forms of punishment, is poised to continue with the passage of such recent legislation as the Sarbanes–Oxley Act. Prompted by a series of high-profile accounting irregularities that resulted in at least a perceived erosion of public trust in corporate oversight, the Sarbanes–Oxley

Act imposes a number of restrictions related to accounting practices and enhances the penalties that may be imposed for violation of securities laws. For example, improperly certifying certain financial statements by an officer of a corporation can result in a fine of as much as $5 million and imprisonment for up to twenty years. The act also created the Public Company Accounting Oversight Board, which oversees auditors of public companies.

These recent developments are simply the latest attempts to address a long-known problem of how to apply the traditional principles of criminal law to corporations in a meaningful and effective way. The basic difficulty is one that was expressed rhetorically by Lord Edward Thurlow in his capacity as the lord chancellor of England: "Did you ever expect a corporation to have a conscience, when it has no soul to be damned, and no body to be kicked?"[52] In the context of the framework described earlier, this question may take a modern form of asking which acts can be imputed to an *actus reus* of the corporation and asking how a mental state of the corporation can be determined to evaluate whether the corporation had the necessary *mens rea*.

These are not easy questions to answer. Fundamentally, the performance of acts and the conception of mental states are specific to individual persons. Should a corporation be liable only when some person has the requisite combination of *mens rea* and *actus reus*? Or is there some way in which acts and thoughts by multiple individuals may be considered in a collective way to reflect the state of the corporation? There is a large body of law that illustrates the difficulties that arise when attempting to grapple with these issues.

At the risk of oversimplifying, what these deliberations essentially conclude is that a corporation may be liable for actions performed by any agent of the corporation, as long as the agent is acting within the scope of her employment or authority and for the benefit of the corporation.[53] These requirements are minimal. They are so broadly applied that an act performed by virtually anyone paid by the corporation has the possibility of being imputed to the corporation itself. It does not matter that the act might be performed primarily for the benefit of the employee herself, as long as there is some incidental intent to benefit the corporation. It also does not matter that the act might not have the sanction of any acknowledged authority within the corporation, potentially making the corporation liable for the actions of rogue employees. The attempt to provide some benefit to the corporation does not even need to succeed, as long as such a benefit was at least one motivation.[54]

This requirement that there be an intent to benefit the corporation is reminiscent of the *mens rea* principle that is also required. But an important distinction needs to be borne in mind. The determination of whether there was an intent to benefit the corporation bears on whether the improper act—the *actus reus*—can be imputed to the corporation. Such an intent is not part of the *mens rea* requirement, which instead remains a separate element that must be satisfied in establishing that a crime has been committed. The mental state that is relevant in assessing the *mens rea* requirement is the mental state defined by the crime.

In the same way that acts by agents of the corporation are imputed to it, the mental states of its agents may also be imputed to it. But the mental state of the corporation may be much more expansive than simply the mental states of the individual agents. The doctrine of *collective knowledge* aggregates the knowledge of the corporation's agents in defining the overall mental state of the corporation. What this means is that the corporation cannot escape liability for a crime by compartmentalizing tasks among its employees. One paradoxical result of this doctrine is that a corporation may be liable for committing a crime—by having imputed to it both the necessary *actus reus* and *mens rea*—even though there is no single employee who is at fault. It is not simply the case that a corporation is liable for the criminal acts of its employees. Rather, such acts represent only a subset of the crimes for which the corporation may have liability.[55]

## 4.  Example: Application of U.S. Sentencing Guidelines to Nanotechnology Crimes

In the United States, the punishment imposed on a corporation that has been found guilty of a crime has, for the last fifteen years, been determined by application of Sentencing Guidelines promulgated by the Sentencing Commission formed through the Sentencing Reform Act. Originally, application of these guidelines in determining punishment was mandatory. But in 2004, the Supreme Court of the United States ruled the mandatory character of the guidelines was unconstitutional.[56] Currently, the guidelines are instead considered to be advisory, and in practice most courts continue to find the advice that they contain to be valuable and impose punishments in accordance with the guidelines. The manner in which the guidelines recommend determining punishment thus remains of considerable importance to corporations.

The punishment recommended for a corporation convicted of a crime is determined by an algorithm that considers a number of different factors that attempt to account for the severity of the crime and the degree to which the organization itself—as opposed to the individual actor—is culpable.[57] The first step in the algorithm is to determine the *offense level* of the crime according to a tabulation of crimes that assign offense levels from 1 to 43. Crimes that are considered to be more serious are assigned higher offense levels.

Some examples of offense levels that might be relevant to nanotechnology companies include violations of regulations dealing with food, drugs, or cosmetics (offense level 6), unlawfully transporting hazardous materials in commerce (offense level 8), providing false information relating to consumer products (offense level 16), endangerment resulting from mishandling hazardous substances (offense level 24), and tampering with consumer products (offense level 25). In some instances, the offense level may vary with circumstances, such as when explosive materials (as might be created by a company supplying nanoenergetic materials) are unlawfully transported. The offense level in such circumstances may be as low as 12 or may be as high as 24 if

**TABLE III.1**

Offense Levels and Base Fines

| Offense Level | Amount | Offense Level | Amount |
|---|---|---|---|
| 6 or less | $5,000 | 22 | $1,200,000 |
| 7 | $7,500 | 23 | $1,600,000 |
| 8 | $10,000 | 24 | $2,100,000 |
| 9 | $15,000 | 25 | $2,800,000 |
| 10 | $20,000 | 26 | $3,700,000 |
| 11 | $30,000 | 27 | $4,800,000 |
| 12 | $40,000 | 28 | $6,300,000 |
| 13 | $60,000 | 29 | $8,100,000 |
| 14 | $85,000 | 30 | $10,500,000 |
| 15 | $125,000 | 31 | $13,500,000 |
| 16 | $175,000 | 32 | $17,500,000 |
| 17 | $250,000 | 33 | $22,000,000 |
| 18 | $350,000 | 34 | $28,500,000 |
| 19 | $500,000 | 35 | $36,000,000 |
| 20 | $650,000 | 36 | $45,500,000 |
| 21 | $910,000 | 37 | $57,500,000 |
| | | 38 or more | $72,500,000 |

there are prior convictions for violent crimes. Similarly, if tampering with consumer products results in a death, the offense level may be increased to the level of a homicide, that is, as high as the maximum of 43 for a first-degree murder if the death was caused intentionally or knowingly.

From the offense level, a base fine amount is determined according to Table III.1. The range of base fines is broad, with the base fine for the most severe crimes being almost 15,000 times as great as the base fine for the least severe crimes.

The base fine is subject to adjustment based on calculation of a *culpability score*. Basically, the culpability score attempts to measure the degree to which the corporation was complicit in the commission of the crime. Such factors as participation in the underlying act by high-level personnel, a history of similar conduct by the corporation, or violation of a judicial order in committing the underlying act tend to increase the culpability score. At the same time, such factors as implementing an effective ethics program and voluntary disclosure of the underlying act to relevant authorities result in a reduction of the culpability score.

The culpability score is then translated into a range of multipliers that are applied to the base fine. The final recommended fine is expected to be within the resulting fine range, with the court examining other factors to set the fine assessed within the final range. At the lowest culpability-score levels, this range of multipliers is 0.05–0.20 and at the highest levels, the range of multipliers is 2.00–4.00. Thus, with the application of these multipliers, the range of potential fines varies by a factor that exceeds 1,000,000.

What all this means for nanotechnology corporations is that diligent efforts to prevent employees from committing crimes that could implicate the corporation are strongly rewarded. The potential exists for there to be a reduction in the base fine by as much as 95 percent if the corporation meets the highest standards set out by the guidelines in deterring crime. And the potential exists for the base fine to be augmented by as much as 300 percent if the opposite is true and the corporation appears to be supporting the criminal activity.

What the guidelines effectively encourage is the adoption of internal policing programs by corporations so that improper acts can be detected and reported. There is a natural tendency for the boards of companies to want to insulate the company from wrongdoing by an employee, particularly when the employee is relatively highly placed in the organization. After all, there is rarely any affirmative duty for people to report their awareness of a criminal act, and as long as someone is not acting as an accessory to the crime, there are no adverse consequences for failing to volunteer information about the crime.

The structure of the sentencing guidelines for organizations actually presents certain interesting dilemmas for the boards of companies. Even though a particular criminal act might have been performed by only a single person, the company itself is acquiring liability for that act. One of the fundamental duties of the board is to protect the interests of the company. While disclosure of the criminal act not only has the potential to implicate a colleague and potential friend of board members but also to produce negative publicity about the company, it is also true that failure to disclose it increases the potential criminal liability being faced by the company.

At first blush then, it would seem that it is in the best interests of the board to disclose the act. In fact, another factor that could impact the decision of the board is the likelihood that the illegal act will be discovered. If the probability of discovery is low, a purely economic calculus might suggest that concealment is the better course of action. This calculus is complicated further when the board recognizes that criminal liability is not the only form of liability that the company may face. There is also the potential for civil liability, and this potential increases if the company discloses.[58] While discretionary factors may also play a role, there is not as direct a benefit in reducing liability from disclosure in the civil context as in the criminal context. From a purely economic perspective, then, the interests of the company are best served by weighing the potential increase in civil liability against the potential decrease in criminal liability that results from disclosure. The dilemma that this creates for companies is a very real one: it seems somehow counterintuitive that the directors of a corporation could be failing in their duties by disclosing criminal activity that they are aware of. But it does appear that, under at least some circumstances, innocent shareholders have a legitimate complaint that the directors have failed to protect their interests.[59]

## 5. International Trends

The previous discussion emphasizes the approach to corporate criminal liability as that doctrine has developed in the United States. One reason for this
emphasis is that the level of liability to corporations and other organizations
for the acts of their employees is probably strongest in the United States.
Indeed, the doctrine of corporate criminal liability remains largely unestab-
lished in many other countries. Another reason for this emphasis is that a
large number of nanotechnology companies are likely to develop as international and transnational entities that have substantial business interests
in the United States. Those corporations that have a presence in the United
States and other jurisdictions that recognize the doctrine would do well to
understand the scope of liability that exists and the mechanisms available to
minimize it.

The international trend appears to be developing toward greater imposition of criminal liability on corporations. The concept is now well accepted
by common-law jurisdictions—the United Kingdom, Canada, Australia, the
United States, and so forth—and many continental European jurisdictions,
notably France and Italy, have adopted criminal-liability provisions for corporations in recent years. In Asia, the doctrine is well established in Japan,
and China has also begun introducing a system of imposing criminal liability on corporations.

While these various countries have different structures and attitudes
towards the idea of, and implementation of, corporate criminal liability, it
is also probably true that their approaches are evolving along lines similar
to those of the United States. There might always be differences of emphasis
and differences in the specifics of implementations, but certain general strategies that governments use will probably evolve similarly as governments
observe the positive and negative effects of different approaches in countries. For example, Japan, which may have the most well-developed system
of corporate criminal liability outside the United States, has imposed duties
on corporate organizations to implement prevention programs that deter
employees from committing crimes within the scope of their employment.
Those who fail to implement such programs in an effective way may suffer
greater liabilities in a manner that bears at least some resemblance to the
approach adopted in the United States.

Concurrent with the evolution of laws in different countries recognizing
criminal liability for corporations will be efforts at harmonizing those laws.
While they still face a number of significant barriers, there are signs that
pressure is being exerted from different directions for various countries to
adopt such provisions.[60] Nanotechnology companies that have multinational
aspects are likely to benefit from such harmonization efforts so that the
effects of internal policing mechanisms are likely to have similar effects in
different countries.

## 6. Prevention and Detection of Crimes Using Nanotechnology

The beginning of the "Criminal Liability" section recalled a recent period of time in which certain criminal acts were effective in causing widespread fear, and drew an analogy between those acts and acts that could potentially be perpetrated using nanotechnology products. There are numerous ways in which the analogy can be criticized, perhaps mostly because it intentionally exploits a widespread impression that the development of nanotechnology will lead inevitably to the production of very harmful structures. This impression has been driven by a number of prominent suggestions that the logical progression of nanotechnology will lead to "molecular assembly." Indeed, the main proponent of such ideas, K. Eric Drexler, is generally credited with having originally coined the term "*nanotechnology*" to refer to such processes, although the term is now completely accepted as referring to a much broader set of technologies.[61]

"*Molecular assembly*" refers specifically to processes that permit extremely precise positioning of atoms and molecules in creating essentially any structure. The pinnacle of such processes would be self-replication by synthetic mechanical nanometer-scale devices in a fashion analogous to biological reproduction. Although this concept of self-replicating "nanobots" has received a good deal of publicity and has been promoted in a number of science-fiction books and movies, it remains subject to significant skepticism within the scientific community. A noteworthy criticism of these ideas was leveled by Richard Smalley, a recipient of the Nobel Prize in chemistry for his contribution to the discovery of fullerenes. In a series of public letters, Drexler and Smalley debated the scientific basis for a molecular assembler, with Smalley claiming that fundamental limitations imposed by the mechanics of chemistry prevented such a device from ever being possible. The exchange ended with a piercing criticism of Drexler and his ideas:

> You and people around you have scared our children. I don't expect you to stop, but I hope others in the chemical community will join with me in turning on the light, and showing our children that, while our future in the real world will be challenging and there are real risks, there will be no such monster as the self-replicating mechanical nanobot of your dreams.[62]

In this spirit, it seems worth concluding a discussion on the relationship of nanotechnology to criminal liability by noting the substantial promise that nanotechnology provides in preventing crimes and in assisting in the investigation of crimes. Nanoscale sensors may find applications in monitoring exposures to toxic gases. Microfluidics devices may be used in forensic applications to identify very small pieces of evidence or may be used in the rapid identification of explosives. Lightweight materials may provide more effective armor for police forces to wear in riot situations. Nanoscale structures may be embedded in currency notes to greatly increase the ability to detect forgeries. Nanotech devices may be used in sophisticated surveillance applications.

These examples provide the merest hint of the potential applications that nanotechnology may find in the detection and prevention of crimes. A recent development in this vein allows this section to come full circle. In September 2006, a group of researchers at Clemson University in South Carolina published results of a study using sugar-coated single-walled carbon nanotubes to aggregate anthrax spores.[63] At one level, this permits a testing methodology to screen suspect powders for the presence of anthrax spores. But at another level, the aggregation that is achieved results in the formation of structures having a size that makes them very difficult to inhale into a person's lungs—the technique may provide a countermeasure to anthrax attacks. In a very real way, this type of research may provide a way to neutralize the sort of fear that resulted from the murder of Robert Stevens.

### Discussion

1. One apparent difference in the previous discussions of civil liability and criminal liability is the level of intent required. A cause of action like negligence does not require any allegation that a harm was inflicted intentionally; this is in marked contrast to the *mens rea* requirement of most crimes. But this difference is an artifact of the way the discussions have been structured. In fact, there is another class of civil causes of action called *"intentional torts"* in which intent of the actor must be proved. They are similar to crimes (and are historically derived from crimes) but are addressed through actions brought by private parties instead of by the state. Discuss the role of such intentional torts in determining the liability of harmful acts caused by employees of a corporation. For example, compare the total criminal and civil liability that results from a harm caused through the negligent release of a harmful nanotechnology product with the liability that results when the product is released intentionally.

2. This section has discussed the ways in which existing criminal laws might be applied to nanotechnology. Do you see any areas in which existing laws seem simply to be inadequate? That is, are there aspects of nanotechnology that will permit new crimes to be committed that are so far outside the conception of current laws that current laws will not apply effectively? In what areas do you think this is true? What sorts of new laws would you propose to address these deficiencies?

3. The effects of punishing corporations include a variety of negative consequences to innocent parties—reductions in shareholder values, the potential loss of jobs as a company's profitability is reduced, and a variety of ripple effects as the financial security of these innocent parties is reduced. It seems easier to justify these consequences in a civil context where compensation is being provided to a person harmed by actions of the corporation's employees. But it seems harder to justify in a criminal action when it is the state that collects

the punishment. Do you think there are sufficient societal justifications for punishing corporations instead of limiting punishment in a personal way to those who directly committed the wrongful behavior? If so, what are those justifications? If not, why do you think the modern trend is for imposition of corporate criminal liability?

4. This section suggests that corporations will engage in an impersonal and mechanical cost assessment of how to respond to discoveries of criminal acts committed by their employees, sometimes determining that concealment of those acts is the most cost-effective response. Do you have any moral qualms about a system of criminal law that promotes this type of calculus? What modifications could you propose to the system to dilute the relevance of such calculations? Do you see any drawbacks with such proposals? For instance, one could prohibit the availability of civil action against a company that discloses criminal activity, but is such an approach consistent with the compensatory goals of the civil system? Is there an intermediate approach that achieves a better balance?

## Notes

1. The novel *The Legend of Bagger Vance* by Steven Pressfield (New York: William Morrow, 1995) is an accessible retelling of the story of *Bhagavad-Gita* in which the battlefield is substituted with a golf course. A movie version of the novel directed by Robert Redford was released in 2000.

2. Excerpt from speech by J. Robert Oppenheimer to Association of Los Alamos Scientists, from *Robert Oppenheimer: Letters and Recollections*, ed. Alice Kimball Smith and Charles Weiner. 1995. Stanford, CA: Stanford University Press.

3. An interesting analysis of the role of the *Bhagavad-Gita* and other Hindu texts in shaping the philosophy of Robert Oppenheimer is provided by Hijiya, James A. 2000. The *Gita* of J. Robert Oppenheimer, *Proceedings of the American Philosophical Society* 144: 123–167.

4. United States Atomic Energy Commission. 1954. In the Matter of J. Robert Oppenheimer: Transcript of Hearing before Personnel Security Board. Washington, DC: Government Printing Office, p. 81.

5. Oppenheimer died of throat cancer on February 19, 1967. His obituary in the *New York Times* on that day captures well the conflict in his psyche, a conflict that is reminiscent of the dual nature of the societal impact that many technological advances can have:

Starting precisely at 5:30 A.M., Mountain War Time, July 16, 1945, J. (for nothing) Robert Oppenheimer lived the remainder of his life in the blinding light and the crepusculine shadow of the world's first man-made atomic explosion, an event for which he was largely responsible.

That sunlike flash illuminated him as a scientific genius, the technocrat of a new age for mankind. At the same time it led to his public disgrace when, in 1954, he was officially described as a security risk to his country and a man with "fundamental defects in his character." Publicly rehabilitated in 1963 by a singular Government honor, this bafflingly complex man nonetheless never fully succeeded in dispelling doubts about his conduct during a crucial period of his life.

6. One of many articles that summarizes research efforts on the application of nanotechnology in medicine is Kubik, T., Bogunia-Kubik, K., and Sugisaka, M. 2005. Nanotechnology on Duty in Medical Applications. *Current Pharmaceutical Biotechnology* 6: 17–33.

7. More precisely, a rebuttable presumption is created that each of the breaching parties caused the harm. Such a presumption permits a plaintiff to satisfy the direct causation element of the negligence cause of action, while maintaining an ability of any of the defendants to offer evidence that he or she was not, in fact, a cause of the harm.

8. For a discussion of nonmonetary remedies, see Discussion question no. 2.

9. In practice, the assignment of such values may well be subject to a variety of issues, but at least the abstract process to be followed is well defined—a circumstance quite different from personal harms.

10. Typically, a further calculation is performed to reduce the difference to a present value, reflecting the fact that the damages award is made all at once and not over a period of time.

11. See Discussion question no. 4 regarding punitive damages.

12. See "The Actual Facts About the McDonald's Coffee Case." http://www.lectlaw.com/files/cur78.htm.

13. *Esteban Ortiz v. Fibreboard Corp.*, 527 U.S. 815 (1999).

14. Studies differ to some degree in their conclusions, although all are consistent with the order of magnitude of the numbers provided in the main text. See Carroll, Stephen J., Hensler, Deborah, Gross, Jennifer, Sloss, Elizabeth M., Schonlau, Matthias, Abrahams, Allan, and Ashwood, J. Scott. 2002. *Asbestos Litigation*. Santa Monica, CA: Rand Corporation; Stiglitz, Joseph E., Orszag, Jonathan M., and Orszag, Peter R. 2002. The Impact of Asbestos Liabilities on Workers in Bankrupt Firms. Research report commissioned by American Insurance Association; White, Michelle J. Summer 2003. Resolving the "Elephantine Mass." *Regulation* 26(2): 48–54.

15. For a general discussion of nanoparticle toxicity, see Howard, C. Vyvyan, and Ikah, December S. K. 2006. Nanotechnology and Nanoparticle Toxicity: A Case for Precaution. In *Nanotechnology: Risk, Ethics, and Law*, ed. Geoffrey Hunt and Michael Mehta, 154–166. Sterling, VA: Earthscan.

16. Id.

17. Some examples of relevant reviews of various studies include Hardman, Ron. 2006. A Toxicologic Review of Quantum Dots: Toxicity Depends on Physicochemical and Environmental Factors. *Environmental Health Perspectives* 114: 165–172; Dreher, Kevin L. 2004. Health and Environmental Impact of Nanotechnology: Toxicological Assessment of Manufactured Nanoparticles. *Toxicological Sciences* 77: 3–5; and Powell, Maria C., and Kanarek, Marty S. 2006. Nanomaterial Health Effects—Part 1: Background and Current Knowledge. *Wisconsin Medical Journal* 105: 16–20.

18. This statement is probably not unfair, but it is worth noting that despite a perception of huge increases in product-liability litigation, there has actually been a recent trend that sees fewer filings of such cases. This did indeed follow a substantial upward trend about twenty years ago. Of interest is the fact that the more recent reduction in filings has been accompanied by an increase in the size of recovery awards. This is probably a result of better identification of meritorious cases by product-liability attorneys in accordance with prevailing laws.

19. *Greenman v. Yuba Power Products*, 59 Cal. 2d 57 (1963).

20. *Soule v. General Motors*, 8 Cal. 4th 548 (1994).

21. Restatement (Third) of Torts: Products Liability, §2(b).

22. In unusual circumstances, it may still be possible to prove that a design is defective even without showing the existence of a reasonable alternative design. This may be done in instances where the design is "manifestly unreasonable" by having an obvious and extremely high degree of risk with no meaningful countervailing utility so that no reasonable person would ever use the product.

23. Restatement (Third) of Torts: Products Liability, §2(a).

24. Restatement (Third) of Torts: Products Liability, §6(d).

25. Uniform Commercial Code (UCC), §2-313.

26. UCC, §2-314.

27. The complete maxim is sometimes said to be *caveat emptor, qui ignorare non debuit quod jus alienum emit* (Let a buyer beware, for he ought not be ignorant of what they are when he buys the rights of another). Hob 99 [Hobart's King's Bench Reports (1613–1625)].

28. Although the doctrines of *caveat emptor* and *caveat venditor* were considered in Roman times, these particular phrases to describe the doctrines appear not actually to have been used. Radin, Max. 1927. *Handbook of Roman Law* §83: 231.

29. Sherman, Charles Phineas. 1937. *Roman Law in the Modern World.* §790, p. 347.

30. UCC, §2-315.

31. The notable exception is Louisiana. Although it has adopted most provisions of the Uniform Commercial Code, it has not adopted Article 2, which governs sales of goods, including warranty provisions.

32. In most legal actions, appeals are only permitted after some "final" judgment has been rendered. It would be too burdensome on the court system if it were possible to appeal every ruling that one party disagreed with. The adversarial nature of most court proceedings all but guarantees that (at least) one party will be dissatisfied with each ruling that a court makes. The decision to certify or to deny certification of a class is one exception to this general rule. Provisions usually exist to allow an appellate court to consider the decision of a lower court on this issue under the right circumstances—for example, when certification would unduly pressure a defendant into settling or when denying certification would act to end the litigation because of impracticalities with pursuing claims individually. See, for example, *Blair v. Equifax Check Services, Inc.*, 181 F.3d 832, 833 (7th Cir. 1999). Appeals that are permitted before a final judgment are referred to as *interlocutory* appeals.

33. See Discussion topic no. 1 for a comparison of opt-out and opt-in forms of class actions.

34. One study found that a median opt-out rate in more than four hundred class actions was on the order of 0.1 percent. Willging, Thomas E., Hooper, Laural L., and Niemic, Robert J. 1996. An Empirical Study of Class Actions in Four Federal District Courts: Final Report to the Advisory Committee on Civil Rules. Federal Justice Center.

35. Id.

36. In the four district courts considered in the Willging study, between 9 and 21 percent of class-action cases (depending on the particular court) had at least one opt-out at the certification stage, but between 36 and 58 percent had at least one opt-out at the settlement stage.

37. An interesting overview of how spin can be used in practical device applications is provided by Das Sarma, Sankar. 2001. Spintronics. *American Scientist* 89: 516–523.

38. Seeger Weiss LLP Announces a Class Action Lawsuit Against NVE Corporation–NVEC. March 7, 2006. Press release, Seeger Weiss LLP.

39. *Joseph Goldstein v. Schering-Plough Corp.*, BC305270; *Christopher Rovere v. Schering-Plough Corp.*, RG03-127026; *Glynis Lowd v. Schering-Plough Corp.*, RG03-128941; *Cristina Williams v. Schering-Plough Corp.*, BC3067444; *Robert Gatson v. Schering-Plough Corp.*, BC310407; *Michael Fong v. Sun Pharmaceuticals Corp. and Playtex Products, Inc.*, BC328875; *Michelle Basch v. Tanning Research Laboratories, Inc.*, BC328758; *Steve Engel v. Neutrogena Corp. and Johnson & Johnson, Inc.*, BC307288; and *William C. Jordan v. Chattem, Inc.*, BC310536.

40. For a thorough discussion of these and other public-policy issues surrounding class actions, see Hensler, Deborah R., Dombey-Moore, Bonnie, Gibbens, Beth, Gross, Jennifer, Moller, Erik K., and Pace, Nicholas M. 2000. *Class Action Dilemmas: Pursuing Public Goals for Private Gain*. Santa Monica, CA: Rand Corporation.

41. Other types of legal services that do not involve litigation are also frequently provided on a fixed-fee basis, with an attorney agreeing to draft a will, perform a real-estate closing, draft a patent application, process an immigration application, and so forth for a predetermined fee, irrespective of the actual effort required.

42. In class actions that do not have common funds—typically those requesting declaratory judgments or injunctive relief—attorney fees may be paid personally by the class members or less commonly by the opposing party.

43. An interesting model of the relationship between fee determinations and incentives for attorneys involved in class actions is presented in Klement, Alon and Neeman, Zvika. 2004. Incentive Structures for Class Action Lawyers. *Journal of Law, Economics, and Organization* 20: 102–124.

44. *Black's Law Dictionary*, 6th ed. 1990. St. Paul, MN: West Publishing.

45. Thompson, Robert B. July 1991. Piercing the Corporate Veil: An Empirical Study. *Cornell Law Journal* 76: 1036.

46. Bush, Larry M., Abrams, Barry H., Beall, Anne, and Johnson, Caroline C. 2001. Index Case of Fatal Inhalational Anthrax Due to Bioterrorism in the United States. *New England Journal of Medicine* 345: 1607–1610.

47. This is not a universal distinction between acts that give rise to criminal and civil liability. There are, for instance, a variety of "intentional" torts that involve the deliberate infliction of harm and that are decidedly civil actions. See Discussion question no. 1.

48. Miziolek, Andrzej W. 2002. AMPTIAC Newsletter 6: 43–48.

49. For purposes of simplicity, the description in this section makes reference to corporations and to corporate criminal liability. But the principles discussed are more generally applicable to any form of organization of a business, including partnerships, associations,

unions, nonprofit organizations, and so on, in addition to corporations.

50. Cohen, Mark A. 1991. Corporate Crime and Punishment: An Update on Sentencing Practice in the Federal Courts, 1988–1990. *Boston University Law Review* 71: 247.

51. See generally Parker, Jeffrey S. 1988. Corporate Criminal Penalties in the Era of Federal Sentencing Reform—Increased Exposure Under the Sentencing Reform Act of 1984. *Corporate Criminal Liability Reporter* 2: 48–57.

52. See, for example, Coffee, John C. Jr. 1981. "No Soul to Damn: No Body to Kick": An Unscandalized Inquiry into the Problem of Corporate Punishment. *Michigan Law Review* 79: 386.

53. *New York Central & Hudson River R.R. v. United States*, 212 U.S. 481 (1909).

54. One example is *United States v. Automated Medical Laboratories, Inc.*, 770 F.2d 399 (4th Cir. 1985).

55. See generally Bajkowski, Sean, and Thompson, Kimberly R. 1997. Corporate Criminal Liability. *American Criminal Law Review* 34: 445.

56. *United States v. Booker*, 543 U.S. 220 (2005).

57. Strictly, the algorithm described here is applicable only to corporations that were not formed primarily for a criminal purpose. Such organizations are instead fined an amount that is "sufficient to divest the organization of all its net assets"—an organizational counterpart to capital punishment.

58. A related point is that even the criminal liability represented by the federal sentencing guidelines is only one form of criminal liability—that imposed by the federal government. There may well be additional criminal liability imposed by states and other jurisdictions that affect the calculus also.

59. The economic considerations may be illustrated in a more quantified way with a simple statistical model. Suppose that the base potential criminal liability being faced by a company for a criminal act committed by an employee is $L$. If the company discloses the act, its liability will be reduced according to a factor $x$ ($<1$) to reflect a lower culpability, but it will be certain to incur a liability of $xL$. If the company does not disclose the act, its potential liability will be increased according to a factor $y$ ($>1$) to reflect a greater culpability. But whether it actually incurs the increased liability of $yL$ depends on whether the act is discovered. Suppose the probability of discovery is $p$ so that the expected liability is $pyL$. Under purely economic considerations, disclosure is indicated when the expected liability for discovery exceeds the potential liability for concealment, that is, when $pyL > xL$ or when $p > x/y$. This result is independent of the

magnitude of the base liability so that the same decision-making process applies for crimes at all levels and disclosure is indicated by a comparison of the probability of discovery to the relevant culpability multipliers.

But now consider the effect of additional liability such as civil liability in a form that does not benefit from the difference in culpability multipliers. Suppose that the base civil liability for the criminal act is $C$. If the company discloses, its total liability is guaranteed to be $C + xL$. If the company does not disclose, its expected total liability is $p (C + yL)$. Applying the same criterion as above, disclosure is indicated when the probability of discovery is $p > (C + xL)/(C + yL)$. In this instance, the actual base liability does remain a consideration, particularly relative to the potential civil liability. And the cutoff probability at which disclosure should be made is always higher than a limited consideration of criminal liability would indicate. Notably, if the potential civil liability is large, $C \gg L$, then the only economically rational point at which disclosure should be made is when the probability of discovery approaches 100 percent.

This model obviously oversimplifies the decision-making process, but illustrates that the impact that the Sentencing Guidelines attempt in encouraging free disclosure of criminal acts by the employees of companies is limited.

60. See, for example, Coffee, J. 1999. Corporate criminal liability: An introduction and comparative survey. In *Criminal Responsibility of Legal and Collective Entities*, ed. by A. Eser, G. Heine, and B. Huber. Freiburg im Breisgau.

61. Eric Drexler's original ideas are presented in his book, *Engines of Creation: The Coming Era of Nanotechnology* (New York: Bantam Doubleday Dell, 1986).

62. Baum, Rudy. 2003. Nanotechnology: Drexler and Smalley make the case for and against "molecular assemblers." *Chemical and Engineering News* 81: 37–42.

63. Want, Haifang, Gu, Lingrong, Lin, Yi, Lu, Fushen, Meziani, Mohammed J., Luo, Pengju G., Wang, Wei, Cao, Li, and Sun, Ya-Ping. 2006. Unique Aggregation of Anthrax (*Bacillus anthracis*) Spores by Sugar-Coated Single-Walled Carbon Nanotubes. *Journal of the American Chemical Society* 128: 13364–13365.

# Conclusion

The preceding chapters have discussed a wide range of legal issues and have illustrated how those issues may bear on nanotechnology. At this point, it is perhaps useful to step back from the details of the legal principles and consider how those principles integrate with other societal issues, particularly with those societal issues discussed in other volumes of the Perspectives in Nanotechnology series. In doing so, a simile that is perhaps apt is one that views the law as a set of tools that may be applied to fashion society in the desired manner. Some of these tools are blunt and their application has the potential to introduce sweeping changes in the way society functions. Others have the capacity for much more precision, being able to target individual issues with greater focus. And—importantly—mechanisms exist that permit new tools to be fashioned.

But how should these tools be used? Should governments take note of the remarkable absence of any significant harms that have arisen involving nanotechnology and allow it develop largely unfettered by interference? Should they instead heed the warnings that some make that there is nonetheless the potential for very serious harms in the future and be very restrictive on the theory that the potential risk warrants extreme caution? Or, as is almost always the correct answer, should some intermediate approach be taken? I do not know the answers to these questions and doubt very much that anyone has a sufficiently detailed understanding of how nanotechnology will develop to know the answers either.

Instead, society will almost certainly rely on the inherent ability of legal mechanisms to act in a self-correcting fashion. These mechanisms have operated in numerous other spheres of human activity over centuries to try to balance the natural competition that arises with any new technology—it has always been the case that the economic and quality-of-life advances made possible by new technologies may be at the expense of a risk that damage will result from unappreciated aspects of the technologies or errors in implementing them. In this way, nanotechnology is no different than various other technologies that have been developed—such as medical technologies, communications technologies, energy technologies, military technologies, and transportation technologies. All of these have benefited from advances that have had profound impacts on the conditions in which human beings live, but have at the same time suffered criticisms that those advances have been at the expense of detrimental impacts on the environment, greater risks to human health, interference with conventional social structures, and so on. And while nanotechnology shares these aspects, it also seems fundamentally different in that it is conceptually broad enough to embrace all of these—and other—classifications of technology.

As more becomes understood about the precise benefits and risks that nanotechnology presents, laws will at times be enforced and enacted in a manner that is found to be so detrimental to the rate of technical progress that demands will be heard to relax the constraining grip of government. And at other times, shocking harms that can be attributed to nanotechnology will undoubtedly occur, causing governments to overreact in the opposite direction by imposing onerous burdens that will stifle further innovation and commercialization of nanotechnology. Over the course of time, the likelihood is that the pendulum will swing both ways multiple times, providing the same sort of convergence on a coherent strategy for balancing issues that the structure of government seeks to achieve.

In attempting to gain insight into specific ways in which governments might use laws to promote and control nanotechnology, though, it is useful to consider the March 2007 report by the Joint Economic Committee of the United States Congress. The committee considered precisely this issue. Its report gives some insight into the early thinking processes of a government body that will need to face these issues. The report speaks both to the concerns raised by nanotechnology that are of specific interest to government and to at least some initial thoughts on the general direction in which it will move in addressing those concerns. Reflecting its understanding of the pace at which nanotechnology is developing, the committee report is provocatively titled "Nanotechnology: The Future is Coming Sooner Than You Think." One of the issues that the report prominently acknowledges is the general disconnect between the current pace of scientific advancement and the rate at which such technology becomes integrated into daily life.[1]

This is actually a fairly significant point. Because the rate at which the scientific and technological information advances in this area is so much greater than the pace at which lawmakers are capable of responding, laws are likely to lag behind development of the most recent nanotech innovations. In some ways, this is no different than other areas of lawmaking, which inherently suffers from the need to take time in deliberating over issues before passing legislation. And this gap between the state of information used in drafting legislation and the state of information when a relevant legal action is initiated is already accommodated by the ability of the judiciary to exercise judgment. It is an important function of the judiciary to discern the underlying intent of the legislature when applying laws to actual facts. This role is in certain respects likely to be greater in actions involving nanotechnology, at least in those specific types of actions where current factual information about the effects of the technology are relevant.

Another important consideration addressed by the Joint Economic Committee's report is the fundamental question that all legislatures ultimately need to confront: "Does nanotechnology represent a danger to society?"[2] In many respects, legislatures would welcome a plain and straightforward answer to this question since it would greatly simplify deciding on a coherent approach to addressing nanotech issues. While everyone should ideally look at the evidence themselves and make an informed judgment about what the level of risk is, my own view is that the report reached roughly the right conclusion. And, at least to me, that is reassuring since it suggests that governments will take a rational approach to the various legal issues that surround nanotechnology. The report acknowledges that there is some evidence of damage to human and animal health from nanoparticles, while at the same time recognizing the conclusion of the highly prestigious U.S. National Academy of Sciences that "'for now there is very little information and data on, or analysis of, [environmental health and safety] impacts related to nanotechnology' and that 'the body of published research addressing the toxicological and environmental effects of engineered nanomaterials is still relatively small.'"[3] In discussing specific illustrations of nanotechnology in this book, I have tried to limit myself to examples that are currently realistic and outside the realm of the fanciful, either positive or negative.

To be sure, this report represents only one of many views that will be considered as various proposals for legislation and regulation are made. And the position taken by the March 2007 report will undoubtedly change as more and better information is acquired. Much of that information makes its way to lawmakers in the United States through the National Nanotechnology Initiative (NNI). The scope of activities of the NNI is broad. Not only does it fund basic nanotechnology research and thereby act similarly to a traditional research funding body, it is involved in a variety of social issues. The NNI itself is organized within the National Science and Technology Council (NSTC), but operates very much as an umbrella organization that coordinates activities related to nanotechnology across the U.S. government. Some twenty-five U.S. government agencies cooperate in an active way with the NNI and the initiative itself has had more than $6.5 billion invested in it since its inception in 2001.

The various social concerns of the NNI largely mirror the social issues discussed in the Perspectives in Nanotechnology series, with emphases being placed on environmental issues, legal issues, ethical issues, and workforce issues, among others. In a very real sense, the integration of social issues in this manner into the mandate of the NNI reflects the way in which nanotechnology is different from other types of technology. In some ways, nanotechnology might be viewed as simply another form of technology that has been developed in a long line of technological developments spanning millennia. But it seems fair to single it out among many of those past technologies for its immense capacity to affect the world. It is capable of impacting a more diverse array of human activities than almost any other class of technology and to do so in a more pervasive way than other technologies have enjoyed. This is true not only from a technical perspective—the way in which nanotechnology permits improvements in almost every sphere of human activity—but also from a social perspective.

The manner in which nanotechnology is likely to affect the lives of human beings is profound. The depth with which it will pervade those lives on a technical level inevitably means that the social aspects of human life will be impacted. Other volumes in the Perspectives in Nanotechnology series have elucidated some of those impacts, illustrating various ways in which social issues have already arisen or are likely to arise. The law is somewhat different. Rather than define a social issue directly, it is perhaps more appropriately viewed as providing a framework within which other social issues can be addressed. It is no less pervasive through society than other social issues; indeed, it must be at least as pervasive in order to provide the structure that is used in addressing those other social issues. This volume of the series has provided a number of illustrations of how that framework is already well suited to addressing the ethical, business, environmental, and other issues that are raised by nanotechnology. It has also illustrated the fluid nature of the law, the way in which the law is structured to be adaptable. This adaptability will prove crucial in addressing the complex issues raised by nanotechnology in a sensible and responsible way.

## Notes

1. Saxton, Jim. 2007. Nanotechnology: The Future is Coming Sooner Than You Think. Joint Economic Committee, United States Congress, p. 6.
2. Id., p. 9.
3. Id., p. 13.

# *Index*

## A

Administrative Procedure Act, 93
AEC, *see* Atomic Energy Commission
Agency action, political and judicial control
      over, 140–147
    discussion, 146–147
    judicial control, 143–146
    political control by executive, 142–143
    political control by legislature, 141–142
Agency for Toxic Substances and Disease
      Registry, 92
Architectural works, 56
Arms Control and Disarmament Agency,
      137
Artificial cells, 71
Artistic works, nanotech creations as, 52–59
    architectural works, 56
    conventional art, 53–54
    Cornell University nanoguitars, 58
    discussion, 59
    literary works, 56–57
    musical works, 58–59
    nanoart, 53
    nanosized sculpture, 54–56
    Osaka Bull, 54
    photonic tweezers, 55
    "Power", 53
    quantum nanocomputer programs, 57
Atomic bomb, detonation of world's first, 134
Atomic Energy Act, 92
Atomic Energy Commission (AEC), 92
Attorney fees, class actions and, 207–209,
      210

## B

Bayh–Dole Act, 83, 85
Biological products, regulation of, 104
Biotechnology inventions, 21, 28
Board of Patent Appeals and Interferences,
      24
Buckminsterfullerene, 43, 44

Buckyball, 44
Bundle of rights, 80
Business organizations, 211–216
    corporate criminal liability, 222–224
    corporate probation, 222
    corporations, 211–212
    discussion, 215–216
    finances, 213
    natural persons, 212
    piercing of corporate veil, 212–214
    Sarbanes–Oxley Act, 222
    sentencing guidelines, 226
    shareholders' derivative actions,
      214–215
    Watergate investigation, 215

## C

CAA, *see* Clean Air Act
Carbon nanotube(s)
    artistic representation of, 46
    electrical-conductivity properties of, 62
CBER, *see* Center for Biologics Evaluation
      and Research
CDC, *see* Centers for Disease Control
CDER, *see* Center for Drug Evaluation and
      Research
CDRH, *see* Center for Devices and
      Radiological Health
Center for Biologics Evaluation and
      Research (CBER), 103, 104
Center for Devices and Radiological Health
      (CDRH), 103, 104–105
Center for Drug Evaluation and Research
      (CDER), 103, 105
Center for Food Safety and Applied
      Nutrition (CFSAN), 103, 106
Center for Veterinary Medicine (CVM),
      103, 106–107
Centers for Disease Control (CDC), 110
CERCLA, *see* Comprehensive
      Environmental Response,
      Compensation, and Liability Act

Economic Espionage Act, 74, 75
Eggshell-skull rule, 165
Electrical inventions, 21
Energy Reorganization Act, 92
Environmental Protection Agency (EPA),
      92, 94
   air pollution simulation software by, 119
   criteria pollutants identified by, 117
   hazardous wastes designated by, 123
   Integrated Risk Information System, 95
   programs, legal basis for, 111
   rule change proposal by, 94
   Toxics Release Inventory, 95
   water-quality standards, 113
Environmental regulations, 109–128
   assessment, 125–126
   Clean Air Act, 117–119
   Clean Water Act, 111–115
   Comprehensive Environmental
         Response, Compensation, and
         Liability Act, 124–125
   discussion, 126–128
   Federal Insecticide, Fungicide, and
         Rodenticide Act, 119–121
   Resource Conservation and Recovery
         Act, 122–124
   Safe Drinking Water Act, 115–116
   Toxic Substances Control Act, 121–122
EPA, *see* Environmental Protection Agency
Ethics, 155
Export(s)
   control, reasons for, 134
   regulation of, 129–140
      classification, 129–131
      discussion, 139–140
      Export Administration Regulations,
            131–135
      fundamental research exclusion,
            135–136
      international traffic in arms
            regulations, 136–137
      Office of Foreign Asset Control,
            138–139
Export Administration Regulations (EAR),
      131–135
   fundamental research, 135
   reasons for export control, 134
Express warranty, 194–196

**F**

Farm Service Agency, 92
FDA, *see* Food and Drug Administration
FDCA, *see* Federal Food, Drug, and
         Cosmetic Act
Federal Food, Drug, and Cosmetic Act
         (FDCA), 102
Federal Insecticide, Fungicide, and
         Rodenticide Act (FIFRA),
         119–121
FIFRA, *see* Federal Insecticide, Fungicide,
         and Rodenticide Act
Food and Drug Administration (FDA), 92,
         102, 107
   Nanotechnology Task Force, 109
   -regulated products, 108
Food products, labeling of, 109
Forest Service, 92
Formal rulemaking, 93
Frankenfood, 100–101
Fuller, R. Buckminster, 44, 45

**G**

Geodesic dome, 43
Golden-rule arguments, 169
Government, control of nanotechnology
         by, 239

**H**

Hazardous substances, definition of, 124
Hazardous waste(s)
   EPA control of, 123
   household waste and, 127
Health Resources and Services
         Administration, 92
Health and safety regulations, 100–109
   Center for Biologics Evaluation and
         Research, 104
   Center for Devices and Radiological
         Health, 104–105
   Center for Drug Evaluation and
         Research, 105
   Center for Food Safety and Applied
         Nutrition, 106